AI REVOLUTION

Navigating the Future of Artificial Intelligence

by

Boyd Alan Sparrow

CONTENTS

INTRODUCTION 7

CHAPTER 1: UNDERSTANDING ARTIFICIAL INTELLIGENCE 11

What is AI? 12

CHAPTER 2: THE EVOLUTION OF AI 20

Early Days of AI 21

Milestones in AI Development 25

CHAPTER 3: MACHINE LEARNING AND DEEP LEARNING 29

Fundamentals of Machine Learning 30

Deep Learning Explained 34

CHAPTER 4: NATURAL LANGUAGE PROCESSING (NLP) 40

Language Modelling 41

Applications of NLP 45

CHAPTER 5: AI IN HEALTHCARE 50

Diagnostic Tools 51

Personalised Medicine 55

CHAPTER 6: AI IN FINANCE 60

Algorithmic Trading 61

Fraud Detection 65

CHAPTER 7: AUTONOMOUS VEHICLES 70

Self-Driving Technology 71

Legal and Ethical Considerations 75

CHAPTER 8: AI IN MANUFACTURING 80

Automation and Robotics 81

Quality Control ____ 84

CHAPTER 9: AI IN ENTERTAINMENT ____ 89

AI-Generated Content ____ 90
Personalised Recommendations ____ 94

CHAPTER 10: AI IN EDUCATION ____ 98

Personalised Learning ____ 99
AI as a Teaching Assistant ____ 102

CHAPTER 11: THE ETHICS OF AI ____ 108

Bias and Fairness ____ 108
Privacy Concerns ____ 112

CHAPTER 12: AI AND EMPLOYMENT ____ 117

Job Displacement ____ 118
New Opportunities ____ 122

CHAPTER 13: AI GOVERNANCE AND POLICY ____ 127

Regulatory Frameworks ____ 128
Public Policy and AI ____ 132

CHAPTER 14: AI AND SOCIETY ____ 137

Social Implications ____ 138
Public Perception ____ 142

CHAPTER 15: THE FUTURE OF AI ____ 147

Emerging Trends ____ 148
Predictions and Speculations ____ 152

CHAPTER 16: AI AND CREATIVITY ____ 156

Art and Design ____ 157
Literature and Writing ____ 161

CHAPTER 17: AI AND CLIMATE CHANGE ____ 165

Environmental Monitoring ____ 166

Renewable Energy 170

CHAPTER 18: AI AND SECURITY 176

Cybersecurity 177

National Security 181

CHAPTER 19: AI IN SPACE EXPLORATION 187

Autonomous Rovers 188

Satellite Data Analysis 192

CHAPTER 20: BUILDING ARTIFICIAL GENERAL INTELLIGENCE (AGI) 197

The Quest for AGI 198

Challenges and Roadblocks 203

CHAPTER 21: HUMAN-AI COLLABORATION 207

Enhancing Human Capabilities 208

Case Studies 213

CHAPTER 22: AI AND PHILOSOPHICAL QUESTIONS 218

The Nature of Intelligence 219

Machine Consciousness 223

CHAPTER 23: INVESTING IN AI STARTUPS 228

Venture Capital Trends 229

Case Studies 233

CHAPTER 24: LEARNING AI: RESOURCES AND TOOLS 239

Online Courses and Tutorials 240

Open Source Projects 244

CHAPTER 25: REAL-WORLD AI APPLICATIONS 249

Case Studies 250

Lessons Learned 254

CONCLUSION 259

Recap of Key Points 260

Future Directions _____ 262

APPENDIX A: APPENDIX _____ 266

GLOSSARY OF TERMS _____ 269

Additional Reading _____ 271

INTRODUCTION

In the annals of technological evolution, artificial intelligence (AI) emerges as a provocative force reshaping our world. The mere mention of AI conjures images of futuristic landscapes, where autonomous machines seamlessly perform tasks, from driving cars to diagnosing diseases. But beyond these futuristic visions lies a profound reality: AI's impact runs deep, influencing everything from the way we communicate to how industries operate. This book seeks to unravel these complexities and reveal the vast potential AI holds for society.

At its core, AI is a testament to human curiosity and ambition. From ancient myths of automata to modern-day algorithms, the journey of AI mirrors our relentless quest to understand and replicate intelligence. This quest is not just about creating so-called intelligent systems; it's about redefining what intelligence itself means. It's fascinating to note how each breakthrough in AI reframes our own perception of human capabilities and potentials.

For the uninitiated, the world of AI might seem like a labyrinth of jargon and technicalities. Yet, it's crucial to remember that behind the complex models and algorithms lie simple yet powerful ideas. These ideas are what propel machines to 'learn' and 'think', to a degree. Today, AI systems draw insights from data in ways that were once the province of human experts, laying the groundwork for transformative applications across various sectors.

However, AI is not merely an array of algorithms waiting to be unleashed upon the world. It's an ecosystem, a convergence of disciplines including computer science, mathematics, neuroscience, and even philosophy. Each field contributes its own perspective, collectively fuelling the progress and ethical considerations of AI. This multidisciplinary nature is both a strength and a challenge, demanding collaboration and continuous dialogue among experts.

The societal implications of AI cannot be overstated. Every day, algorithms influence decisions in realms as diverse as healthcare, entertainment, and finance. But as we marvel at AI's capabilities, we must also pause to reflect on the broader implications. How will AI shift power dynamics? What will be its impact on employment? And how will we navigate the ethical conundrums it poses? These are not merely academic questions but urgent societal challenges.

Drawing inspiration from visionary thinkers, this book will explore AI's potential to both uplift and disrupt. For tech enthusiasts and professionals, understanding the nuances of AI is crucial. This knowledge not only illuminates current trends but also prepares us for what's ahead. As AI continues to evolve, so too must our frameworks for understanding and engaging with it. The balance between innovation and ethical integrity will be paramount as we forge this new frontier.

Amidst the myriad of possibilities, AI has also sparked plentiful debates, not just about ethics and governance, but about the very nature of consciousness and creativity. Can machines ever be truly creative or conscious? While these philosophical questions may seem abstract, they touch on essential aspects of humanity. As we journey deeper into the era of AI, our understanding of these concepts may shift, prompting us to redefine the boundaries of what it means to be intelligent.

Embarking on the exploration of AI is akin to embarking on a journey through uncharted territory. Curiosity, innovation, and a touch of scepticism are essential companions. As we delve into the chapters that follow, we'll uncover the many ways AI is being harnessed to address challenges and open new avenues for progress. Whether it's transforming industries, enhancing our daily lives, or challenging us to think differently, AI stands as a powerful testament to human ingenuity.

But let's not jump ahead too much. The story of AI is not set in stone; it's a narrative continually evolving, influenced by discoveries, societal shifts, and the ever-changing nature of human needs and desires. This complexity only amplifies the necessity for thoughtful dialogue and responsible development as we innovate.

In bringing this comprehensive exploration to you, our hope is to provide both a landscape overview and a deep dive into pivotal aspects of AI. Through a mix of storytelling, analysis, and forecasts, we aim to equip you with the insights needed to appreciate the multifaceted role AI plays in our world—and in our future. Dive in, explore, and discover the intersection of technology and human ambition.

CHAPTER 1: UNDERSTANDING ARTIFICIAL INTELLIGENCE

Artificial Intelligence, often abbreviated as AI, sits at the forefront of today's technological revolution. It's not just a single technology, but a complex web of innovations that mimic human intelligence's core attributes: learning, reasoning, and self-correction. This field, which once belonged to the realm of science fiction, now permeates various aspects of daily life, from voice-activated assistants to advanced data analysis tools. Understanding AI begins with unpicking its foundational concepts and tracing its roots back through history, from ancient myths of automatons to Alan Turing's early theories. The magic of AI lies in its ability to process vast amounts of information at lightning speed, uncovering patterns and insights that escape the human eye. As technology enthusiasts and professionals, exploring AI means stepping into a world of endless possibilities, where practical application and ethical considerations walk hand in hand. This journey is as much about understanding machines as it is about understanding ourselves and the future we are collectively building.

What is AI?

In the grand tapestry of technological evolution, artificial intelligence (AI) stands out as both a marvel and a mystery. At its core, AI is about creating systems that can mimic human intelligence, albeit in specific, narrow domains. In some ways, it's about engineering machines to think, learn, and adapt—a feat once considered the realm of science fiction. By harnessing vast datasets and employing complex algorithms, AI systems are designed to identify patterns and make decisions, often surpassing human efficiency and accuracy. Whether it's through virtual assistants that anticipate your needs before you've articulated them or algorithms that transform raw data into strategic insights, AI is reshaping our world in subtle yet profound ways. Its potential is vast and its possibilities expansive, creating a future where man's ingenuity is seamlessly woven with machine precision.

Definitions and Key Concepts sit at the heart of understanding artificial intelligence, especially when attempting to unravel the intricacies of what AI truly is. At its core, AI refers to the simulation of human intelligence by machines, particularly computer systems. This simulation encompasses various capacities such as learning, reasoning, self-correction, and, perhaps most intriguingly, adaptation.

AI is not a monolith but a spectrum of disciplines and technologies, encompassing fields like cognitive science, computer science, and even a dash of philosophy. It's about imbuing machines with the ability to perform tasks that would typically require human cognition, from recognising speech to solving complex problems and making decisions. This isn't about single, standalone systems but rather interconnected processes that together enable machines to operate intelligently.

One of the fundamental key concepts within AI is **machine learning**, which is the art of systems automatically improving from experience without being explicitly programmed. Machine learning forms a foundational block upon which many AI processes are built. From predictive analytics to recommendation systems, machine learning algorithms model data and then use it to predict future actions or outcomes.

At a deeper level, machine learning branches into **supervised learning** and **unsupervised learning**, each with unique applications. Supervised learning uses labelled datasets to train algorithms that can classify data or predict outcomes accurately, while unsupervised learning explores data without predefined labels, allowing the algorithm to identify patterns or groupings.

Then there's **deep learning**, a more ambitious subset of machine learning, inspired by the neural networks of the human brain. Deep learning employs layers of algorithms known as neural networks to sift through complex data, enabling applications like image and speech recognition, which come strikingly close to human capabilities. The depth of these neural networks, with their many processing layers, grants machines unprecedented autonomy in recognising patterns and interpreting data.

The term **neural network** often conjures images of biological brains, which is not far from reality, as these networks mimic how human neurons fire. In AI, they serve as the architecture for deep learning models. Neural networks are composed of nodes, analogous to neurons, organised in layers - input, hidden, and output layers. Each node processes input and passes the result to subsequent layers to decipher complex data patterns.

Of course, understanding AI wouldn't be complete without delving into **natural language processing** (NLP), which equips machines with the ability to comprehend and respond to human language. NLP is at the core of various AI operations, such as chatbots, virtual assistants, and other technologies that allow computers to interact seamlessly with humans in a linguistically natural way.

The transformative power of AI stems not only from these technical definitions but from its practical application across various realms - healthcare, finance, transportation, and so forth. It's the blend of these technologies that results in the incredible capabilities we see today, like personalisation in marketing or predictive maintenance in manufacturing.

Moreover, there's the idea of **artificial general intelligence (AGI)**, a key concept that ignites much debate and speculation. Unlike narrow AI, which is designed for specific tasks, AGI aspires to perform any intellectual task a human can do, potentially surpassing human intelligence itself. It's a horizon that pushes ethical, philosophical, and technical boundaries, sparking conversations about the nature of intelligence and consciousness.

A vital concept intertwined with AI is that of **algorithms**, the step-by-step procedures or formulas for solving problems. Algorithms are the invisible hands guiding AI systems, helping them navigate through oceans of data to extract meaningful insights. Their sophistication can vary drastically, from simple linear regressions to complex models like deep learning networks.

Data, too, plays a crucial role. It's the lifeblood of AI, fueling algorithms and allowing them to learn and adapt. The massive proliferation of data in the digital age has been a significant factor in AI's rise, providing the raw material to train complex models. Without data, AI systems would be akin to cars without fuel, unable to perform the tasks they're designed for.

Understanding AI also requires appreciation of the **feedback loop**, a fundamental process by which systems learn. Feedback loops allow AI models to refine themselves based on the results they produce. With each iteration, the model becomes more attuned to making accurate predictions, acting as a cornerstone of the learning process in AI.

This discussion would be incomplete without mentioning **ethics**, a key concern shaping the development and deployment of AI technologies. As AI systems infiltrate every facet of life, ethical considerations regarding privacy, bias, and accountability become paramount. The integration of ethical guidelines into AI processes ensures these powerful tools serve humanity responsibly.

Lastly, there's the notion of **automation**, a byproduct of AI that has revolutionised industries. Automation involves the use of AI to perform tasks that humans once completed manually. While it boosts efficiency and reduces human error, it also raises questions about job displacement and the future of work.

The definitions and key concepts of AI encompass a tapestry of ideas and technologies that together paint a picture of a future imbued with intelligent machines. Understanding these basics is crucial for navigating the rapidly evolving landscape of AI, where the only constant is change and the potential is boundless.

Historical Background The origins of Artificial Intelligence (AI) are deeply entwined with humanity's own intellectual journey. For centuries, we've been fascinated by the idea of creating machines that can think and act like us. The concept harks back to ancient mythologies, such as Talos, the bronze automaton from Greek legend, who was fashioned not as a sentient being but as a perfect executor of orders. Ancient lore across cultures often imagined forms of non-human intelligence, but it was with the advent of modern science and philosophy that these ideas began to crystallise into something resembling today's understanding of AI.

During the Enlightenment, thinkers began to dissect the mechanisms of thought itself, laying the groundwork for later developments in AI. René Descartes famously posited the difference between humans and machines, suggesting that only humans possess the faculties of reason and language. Little did he know that centuries later, thinkers would challenge this very notion, paving the way for the blooming of AI research.

Fast forward to the 20th century, when the field of computer science began to surface. It was here that Alan Turing emerged as a pivotal figure. During the 1930s and 40s, Turing developed foundational theories of computation, producing the concept of a "universal machine". This theoretical machine laid the groundwork for the digital computers we use today. His famous question, "Can machines think?", published in 1950, initiated a revolution that set scientists and philosophers on a quest to emulate human intelligence in machines.

The official birth of AI as a field of study is often pegged to the 1956 Dartmouth Conference, organised by John McCarthy, Marvin Minsky, Nathaniel Rochester, and Claude Shannon. It was here that the term "Artificial Intelligence" was coined, marking the start of a dedicated pursuit to create machines equipped with human-like intelligence. This conference signalled the marriage of theory with ambition, bringing together some of the brightest minds to lay down the first bricks of this vast and ambitious domain.

In these early days, optimism knew few bounds. Researchers envisioned a world where machines would soon match the intuitive prowess of the human mind. Projects such as McCarthy's LISP programming language and the Logic Theorist by Allen Newell and Herbert A. Simon, which could mimic human problem-solving, notched early successes. These advances created a wave of enthusiasm as grant funding and research support surged.

However, the burgeoning enthusiasm was soon met with the harsh realities of technical limitations and unfulfilled promises. By the 1970s, the field faced its first "AI winter" as funding dried up and projects stalled. The expectations had outpaced the maturity of existing technology, breeding scepticism about the timeline for realising human-level machine intelligence. However, this pause enabled researchers to refine their approaches, teaching valuable lessons that would fuel future breakthroughs.

The narrative of AI is interspersed with cycles of breakthroughs and setbacks. The 1980s and 90s saw periods of resurgence, driven by advances in computational power and new methodologies. Expert systems, capable of simulating decision-making in specialised fields, emerged during this time. Although these systems marked significant progress, they also underscored the limitations of rule-based AI, leading to another period of subdued expectations.

Then came a renaissance. The advent of machine learning and neural networks in the late 90s and early 2000s, inspired by the architecture of the human brain, breathed new life into the field. Techniques like backpropagation and increased data availability began to unlock new capabilities, moving AI research beyond the confines of static algorithms to a more dynamic, adaptive framework.

As computing power grew exponentially and data became the lifeblood of the digital age, AI began to infiltrate everyday life in the form of personal assistants, recommendation systems, and more. Machine learning emerged as a pivotal tool, allowing systems to learn patterns from data without explicit programming. This paradigm shift expanded AI's reach into numerous industries, setting the stage for the innovations we witness today.

The historical journey of AI has been a tapestry woven with threads of human curiosity, ambition, and relentless pursuit of knowledge. While the idea of intelligent machines originated in myth and imagination, we've transformed that narrative into a scientifically grounded quest for understanding intelligence itself. As we stand on the shoulders of these intellectual giants, tracing back through centuries of contemplation and decades of computational innovation, AI continues to be both a mirror and a challenge to our understanding of what it means to think and act intelligently.

CHAPTER 2: THE EVOLUTION OF AI

Tracing back to its roots, the evolution of AI has been a rollercoaster of innovation and revelation. From the birth of early computational theories by pioneers like Alan Turing to the sophisticated algorithms that underlie today's intelligent systems, AI's journey is a testament to human ingenuity and perseverance. Initially confined to the realms of theoretical musings and simple machines that could perform basic tasks, AI has now burgeoned into a complex web of technologies that permeate nearly every facet of modern life. This progression hasn't been without its hurdles; early ambitions often outpaced technological capabilities, necessitating periods of introspection and recalibration within the AI community. Yet, each challenge surmounted has only served to fuel the fire of human curiosity, pushing boundaries and elevating the possibilities of what machines can achieve. Central to this narrative are the milestone achievements that redefined expectations—like the triumph of machine learning and the dawn of neural networks—each marking a significant pivot in the evolution of AI. As we stand on the cusp of further advancements, it's clear that the story of AI is far from over, with future chapters promising to be as groundbreaking as the ones we've already turned.

Early Days of AI

The early days of artificial intelligence were marked by a heady mix of aspiration and experimentation. It was a time when pioneering figures, visionaries like Alan Turing, dared to dream of machines that could think, sparking an intellectual revolution. Armed mostly with curiosity and a handful of solid theories, these early AI enthusiasts embarked on a journey that bridged philosophy and engineering. In makeshift labs and dimly lit university halls, they conjured the first algorithms, laying down the bedrock for what would become one of the most transformative technologies of the modern age. This period was not so much about polished products, but rather about grappling with the very nature of intelligence itself. The fledgling ideas conceived in this era served as the prologue to a story of technological wonder that continues to evolve today, unfurling into realms those early dreamers could scarcely have imagined.

From Turing to Today's Technologies In the grand tapestry of artificial intelligence, Alan Turing holds a thread of profound significance. His contributions during the 20th century laid the groundwork for much of what we consider AI today. Turing, an English mathematician and logician, is often heralded as the father of computer science. His groundbreaking 1936 paper introduced the concept of a theoretical machine that could simulate any other machine's logic, sparking the curiosity and creativity of countless researchers.

But Turing's most memorable legacy in AI is perhaps the "Turing Test", a philosophical and practical benchmark for a machine's ability to exhibit intelligent behaviour indistinguishable from that of a human. This concept fascinated scientists and ethicists alike, posing questions about cognition, consciousness, and the elusive definition of intelligence. In a sense, Turing set a challenge for future generations—to create machines that think.

Fast forward to the post-World War II era, when the ambition to bring machines to life took a more tangible form. The creation of some of the first programmable computers during the 1940s and 1950s marked the transition from theory to practice. The Manchester Baby, the world's first stored-program computer, switched on in 1948, offering a glimpse into the future of interconnected networks and intelligent systems.

The 1950s and 1960s were decades of optimism and experimentation in the AI landscape. Researchers started putting Turing's ideas into action with varying success. The Dartmouth Conference in 1956 is often cited as the official birth of AI, gathering brilliant minds whose discussions and research proposals effectively launched the field. Names like John McCarthy, Marvin Minsky, and Herbert Simon became synonymous with pioneering efforts in artificial intelligence, helping to forge the path that Turing had envisioned.

During these early days, symbolic AI emerged as a predominant paradigm. Researchers sought to encode human knowledge in symbols and rules that machines could parse. This symbolic approach led to advancements like the Logic Theorist, an early AI program capable of proving mathematical theorems. Excitement bubbled as possibilities seemed endless, yet challenges soon emerged—complexity and scalability thwarting simpler symbolic methods.

The narrative of AI's evolution took a dramatic turn with the advent of machine learning in the 1980s. This paradigm shift moved away from hardcoded, rule-based systems to embracing data-driven models capable of learning from experience. Turing's logical foundations still applied, but their expression in algorithms evolved. Researchers explored neural networks, inspired by the human brain, allowing computers to recognize patterns, adapt, and self-correct.

As we entered the 21st century, breakthroughs in hardware, particularly the exponential growth of computing power described by Moore's Law, started to supercharge AI's capabilities. With more power at their disposal, researchers could train more complex models on troves of data—something Turing's era couldn't have fathomed. The Internet and digital revolution provided a wealth of data that allowed machine learning algorithms to thrive, ushering in a new age of technological possibility.

What's truly inspiring about this journey from Turing to today's technologies is the democratisation of AI research and development. Open-source software frameworks like

TensorFlow and PyTorch have made it feasible for a vast array of people to experiment with AI, once a realm reserved for those with access to supercomputers. Cloud computing further breaks down barriers, allowing even small enterprises and solo innovators to contribute to AI advancements.

Today, AI applications permeate nearly every facet of modern life with a range of sophistication that might have once seemed like science fiction. Autonomous vehicles, voice assistants, and real-time translation services are just some marvels born from foundational ideas Turing set in motion. His vision prompts each new generation to ponder the nature of intelligence, challenging us to redefine and reach for new heights.

Yet, amidst these achievements, Turing's philosophical queries remain tantalizingly open-ended. As AI systems grow more advanced, the quest for meaningful human-computer interaction intensifies. Conversations about the ethics of AI, its role in society, and the boundaries of its capability reflect the same conceptual debates sparked by the original Turing Test. The journey from Turing's era to today's technologies is not just about computational prowess—it's about exploring what it means to be intelligent.

AI's story is far from over, and in many ways, it's just beginning. Turing's ethos continues to inspire researchers to bridge the gap between theoretical potential and real-world application. As we continue to explore this remarkable frontier, Turing's contributions remain a beacon, guiding discourse and innovation towards a future where machines might coexist harmoniously with the complexities of human life.

Milestones in AI Development

The journey of artificial intelligence from mere concept to transformative technology is marked by a series of pivotal milestones. Each advancement has not only pushed the boundaries of what's possible but has also challenged our understanding of intelligence itself. It's a story of curiosity, innovation, and sometimes unexpected breakthroughs. Let's delve into some key moments that have shaped the field of AI as we know it.

One of the earliest sparks in AI's development was the proposal of the Turing Test by Alan Turing in 1950. Turing's paper, "Computing Machinery and Intelligence," questioned whether machines could think, positing a test where a human evaluator would distinguish between a human and a machine based on their responses to questions. This idea laid the groundwork for contemplating AI's potential, and while still debated, the Turing Test remains a fundamental concept in AI philosophy. Turing's vision prompted a surge of theoretical exploration into the possibilities of thinking machines.

Fast forward to the mid-20th century, and you'll find that AI was undergoing rapid transformation. In the summer of 1956, the Dartmouth Conference officially marked the birth of AI as a field of study. Key figures like John McCarthy, Marvin Minsky, Nathaniel Rochester, and Claude Shannon convened to ponder the potential of creating "machines that learn," coining the term "artificial intelligence". This event set the stage for AI to emerge as a legitimate academic discipline. From then on, research and interest in AI began to blossom worldwide.

Throughout the 1960s and 70s, AI research alternated between periods of optimism and 'AI winters', where progress stalled and funding dried up. However, during these fluctuating times, symbolic AI and expert systems came to the forefront. Expert systems used knowledge-based approaches to solve specific industry problems, heralding a new era where machines could seemingly imitate human expertise in narrow domains. This pushed the boundaries of what was considered plausible for AI applications.

A significant breakthrough came during the late 1990s, when Deep Blue, a chess-playing computer developed by IBM, defeated world champion Garry Kasparov. This victory in 1997 was more than just a triumph in gaming—it was a clear illustration of how far AI had come. Deep Blue demonstrated the potential for machines to perform complex tasks at a superhuman level, inspiring further investigations into the possibilities of AI across different fields.

As the new millennium inched closer, the advent of the internet and accelerated computing power brought AI innovations steadily into everyday life. Machine learning, a subset of AI, began to gain attention due to its ability to 'learn' from data. Tools like Google's PageRank algorithm emerged, showcasing AI's capability to transform information retrieval and internet navigation. Around this time, the focus began shifting from rule-based systems to data-driven models, paving the way for more sophisticated techniques.

The 2010s saw a renaissance in AI with the rise of deep learning, a field that revolutionised machine learning capabilities. Deep learning models, inspired by the structure of the human brain, allowed computers to recognise patterns and make decisions with unprecedented accuracy. A hallmark moment came in 2012 when a deep learning algorithm won the ImageNet competition with a remarkable error rate of just 16%, a breakthrough that made it clear that AI was entering a new era.

Moreover, the AlphaGo program developed by DeepMind in 2016 became a significant AI milestone. By defeating world champion Go player Lee Sedol, AlphaGo showcased the unmatched strategic depth possible with advanced AI. Go, with its complexity and intuition-heavy strategy, was thought to be beyond the reach of conventional AI. AlphaGo's success highlighted not only technical prowess but also a shift in how AI could achieve feats thought largely impossible due to the complexity and nuance involved.

AI has also made significant strides in natural language processing (NLP), enabling computers to understand, interpret, and generate human language. Innovations like OpenAI's GPT-3 and Google's BERT model have transformed NLP by making interactions between humans and machines more seamless and natural. These advancements have paved the way for virtual assistants, chatbots, and automated translation services, making AI a crucial part of everyday communication and information processing.

Beyond specific algorithmic developments, AI milestones are also seen in the evolution of hardware that powers these algorithms. Specialised hardware like GPUs and TPUs have significantly sped up AI computations, allowing for more complex models to be trained in less time. This hardware revolution has supported AI's growth, enabling it to tackle real-world problems more effectively and efficiently.

In recent years, AI has reached into realms like healthcare, finance, and environmental science, showcasing its potential to tackle unique and complex problems. For instance, AI-driven diagnostic tools that use imaging data have started to rival human experts, ushering in a new age of medical diagnostics. Similarly, AI's ability to process vast amounts of financial data faster than any human has opened new strategies in algorithmic trading and fraud detection.

These milestones in AI development reflect a trajectory that is both inspiring and thought-provoking. The journey of AI continues to unfold, with each new discovery challenging our assumptions and expanding the horizon of what's possible. The field stands on the shoulders of these pivotal moments, each a testament to human ingenuity and our relentless pursuit of understanding intelligence in all its forms.

CHAPTER 3: MACHINE LEARNING AND DEEP LEARNING

As we transition from the broad strokes of artificial intelligence's evolution, we delve into the heart of its current powerhouse: machine learning and deep learning. These two pillars mark a significant departure from the rule-based systems of old, replacing them with models that learn from data. Machine learning, with its vast array of algorithms, serves as the toolkit for pattern recognition and predictive analytics. From recommendation systems that hint at your next binge-watch to algorithms that trade billions on the stock market, its impact is profound and pervasive. Deep learning, often described as a subset of machine learning, dives deeper into data through layered neural networks, mimicking the human brain's structure in a quest to unravel complex patterns. It powers advances from voice-activated assistants to autonomous vehicles navigating bustling city streets. The dynamism of these fields lies in their continuous evolution, forever chasing the tantalising prospect of machines that learn faster and smarter, transforming industries in ways we might only have glimpsed in science fiction.

Fundamentals of Machine Learning

Machine learning isn't just a buzzword—it's the powerful engine driving the transformation of data into actionable insight, unlocking possibilities that were once purely the realm of science fiction. At its core, machine learning is all about teaching computers to learn from experience, much like we do. While traditional programming requires explicit instructions to perform tasks, machine learning models find patterns and make decisions based on data. This shift from explicit coding to a learning approach has opened endless vistas in technology, allowing machines to adapt and improve over time. Think of it as giving computers a brain to work with, rather than a list of instructions. This dynamic branch of artificial intelligence has a leg up in various domains because of its versatility, from recommending the next song in your playlist to predicting stock market trends. This adaptive capability challenges us to reimagine what machines can achieve, setting the stage for a world where life and tech are more symbiotic than ever. As we delve deeper into algorithms and models, the threads of innovation in machine learning weave new possibilities, encouraging us to ponder not just the capabilities we can harness today, but the untapped potential yet to be explored.

Algorithms and Models are at the heart of machine learning, shaping how machines can perceive and interpret the world. Picture algorithms as the meticulous recipes and models as the well-trained chefs; together, they form the gourmet kitchen of artificial intelligence. Without them, the promise of AI would remain a theoretical dream rather than the transformative force we witness today.

The diversity of algorithms in machine learning is astounding, mirroring the diversity of challenges they aim to tackle. From simple linear regression models that predict trends with a straight line to complex neural networks that decipher the hidden patterns in enormous datasets, algorithms guide machines in learning from data. The process often begins with data collection and preprocessing, essential for feeding clean and relevant information into our models, just like gathering the finest ingredients for a dish. This initial phase is crucial, for even the most sophisticated algorithms can flounder when fed unreliable data.

Once our data is ready, we enter the realm of training. Here, algorithms adjust their internal parameters to minimize errors in prediction. This training period is akin to the rigorous preparation of a chef honing their craft; the model, through trial-and-error, gradually learns to replicate outputs with increasing accuracy. Take, for instance, supervised learning models, where algorithms are provided with labeled data to learn from known input-output mappings. This contrasts with unsupervised learning, where models must discern patterns from data without explicit guidance. Both strategies have their unique applications and offer insights into different aspects of data exploration and analysis.

In supervised learning, algorithms like decision trees, support vector machines, and k-nearest neighbors showcase their prowess. Decision trees, resembling flowcharts, segment data based on decision rules, making them straightforward and interpretable. On the other hand, support vector machines excel in high-dimensional spaces, ensuring maximum margin separation between data classes. K-nearest neighbors pivot on the assumption that similar data points exist in close proximity, predicting outputs based on the majority label of the nearest examples. Each of these models has its domain of expertise and offers unique ways to solve problems.

Unsupervised learning brings algorithms like clustering and association rule learning to the forefront. Clustering algorithms, such as k-means and hierarchical clustering, aim to discover inherent groupings within the data. Imagine organising a library; clusters define categories that render otherwise chaotic collections navigable. Association rule learning delves into finding interesting relations between variables, often employed in market basket analysis to identify common purchasing patterns. This category of models thrives on uncovering hidden structures, aimed at generating insights without predefined labels.

In recent years, a novel subset of machine learning known as deep learning has captured the imagination of researchers and the public alike. At the core of deep learning are neural networks, inspired by the intricate workings of the human brain. These networks consist of layers of interconnected nodes where each connection is assigned a weight that signifies its importance. Through several propagation cycles, these networks adjust the weights to minimize error, evolving to perform tasks like image recognition and language translation with astounding success. Convolutional neural networks, a type of deep learning model, have revolutionised computer vision. Their ability to automatically detect features in images has facilitated advances in autonomous driving and medical diagnostics.

Alongside algorithms are models that enable transfer learning—the concept where pre-trained models, with knowledge garnered from one task, are adapted to new but related tasks. It's akin to a multitasking chef who, having mastered the art of soufflés, easily transitions to baking meringue pies. In practice, transfer learning significantly reduces training time and improves performance on tasks with limited labeled data.

However, our journey with algorithms and models isn't always straightforward. There are challenges such as overfitting, where a model learns too well from its training data, capturing noise instead of underlying patterns. Techniques like regularisation are employed to mitigate this by adding a penalty term to the loss function, ultimately enhancing the model's ability to generalize to new data. Furthermore, hyperparameter tuning—adjusting parameters that are not updated during training—can dramatically influence a model's performance, requiring careful selection and validation.

Evaluating the success of machine learning models involves metrics tailored to their specific tasks. For classification problems, accuracy, precision, recall, and the F1-score are routinely employed to measure performance. Regression tasks might rely on metrics like mean absolute error and root mean square error. These metrics provide quantitative evidence of how effectively a model has learned from the data and is a crucial step in the iterative process of machine learning model development.

As algorithms and models evolve, they transform industries, inspiring solutions once relegated to the realm of science fiction. They bring us one step closer to unraveling the intricacies of human intelligence by expanding the frontiers of what machines can accomplish. As we continue to explore these capabilities, the synergy between algorithms and models promises to redefine how we engage with technology, offering a profound glimpse into the future of machine learning and its boundless potential.

Deep Learning Explained

Deep learning stands at the forefront of artificial intelligence, transforming how machines interpret complex data and make decisions akin to human reasoning. At its heart lies the intricate web of neural networks, a digital imitation of the brain's architecture, where layers of interconnected nodes process information to uncover patterns and insights previously thought to be the domain of science fiction. This powerful computational approach enables the creation of everything from breathtaking images to lifelike speech synthesis, capturing the imagination and driving innovation across industries. Balancing complexity and accessibility, deep learning doesn't just mimic brain functions; it transcends them, pushing the boundaries of what we understand as intelligence. The real magic happens when these networks learn and evolve, adjusting their weights and biases in ways we've only started to comprehend, offering a glimpse into a future where machines uniquely augment human potential and creativity.

Neural Networks and Beyond delve into a realm of technological wizardry that has transformed entire industries and reshaped our very understanding of intelligence. Embarking on a journey through the intricate structures of neural networks, you'll discover how layers of artificial neurons power everything from simple pattern recognition to complex decision-making systems. Often inspired by the human brain, these networks don't just mimic intelligence but evolve to surpass the boundaries of traditional computing, offering a glimpse into the limitless horizons of AI's future possibilities.

The effectiveness of deep learning can be primarily attributed to the architecture of neural networks. These networks, composed of a series of interconnected nodes, are the very essence of deep learning. By simulating the neural pathways of the human brain, they enable machines to learn from vast amounts of data. Each node, or artificial neuron, processes input data and passes it through layers, known as hidden layers, to produce an output. This intricate dance of data is what allows neural networks to perform tasks such as identifying images, recognising speech, and even generating coherent language.

The evolution of neural networks has seen significant leaps, especially with the advent of advanced techniques such as convolutional and recurrent neural networks. Convolutional neural networks (CNNs) bring a unique might in image processing, capitalising on the principle of local connectivity and weight sharing, making them ideal for recognizing patterns in pictures. Meanwhile, recurrent neural networks (RNNs) excel in processing sequences of data such as time series or spoken language owing to their inherent ability to maintain context through loops in their architecture. These developments mark just the beginning of the vast potential neural networks hold.

Jordan and Elman networks made their mark in the early foundations, yet, today's attention mechanisms and transformers have overshadowed them like never before. Attention mechanisms revolutionised the way we approach neural networks by allowing models to weigh the importance of different inputs dynamically. This breakthrough is particularly instrumental in natural language processing, opening avenues for transformer models like BERT and GPT, which have redefined language understanding and generation through their massive scale and efficiency.

As we gaze into the horizon of neural networks, the convergence of these models with other areas of AI stands as a testament to their versatility. Take reinforcement learning, for instance; when coupled with deep learning, it gives birth to deep reinforcement learning. This amalgamation empowers models to make decisions through trial and error, akin to how creatures evolve instincts. The feats achieved by these systems span from mastering games that were once considered the domain of human expertise to navigating complex robotic tasks without explicit programming.

Quantisation and pruning methods make models more efficient by reducing the network's size and computational requirements. This innovation is crucial for deploying AI models on devices with limited resources, such as smartphones or IoT devices without sacrificing performance. Moreover, these techniques contribute to a new wave of "tiny AI", where the power of neural networks is condensed into miniature yet mighty forms capable of operating independently on edge devices.

The advent of generative neural networks opened the door to a world of creativity that artificial intelligence can either assist or autonomously undertake. Generative Adversarial Networks (GANs) create art, music, and more by pitting two networks against each other—one generating creative content and the other criticising it. These networks have grown into tools that redefine artistic boundaries, allowing machines to not just replicate but innovate, serving as collaborators to human artists in the creation of novel, previously inconceivable forms.

Yet, neural networks—and AI as a whole—pose critical challenges and ethical concerns. The immense data requirements and energy consumption of extensive networks bring environmental and economic considerations that can't be ignored. As these models become ever more sophisticated, the imperative of responsible and fair use grows stronger. Bias in AI systems and data security stand as cautionary aspects that experts continually address to ensure that AI grows in a direction that benefits and respects all of humanity.

In understanding neural networks, one can't overlook the profound community efforts that have driven innovation. Open-source platforms and collaborative research initiatives have been pivotal in accelerating development in this field. They've democratized access to cutting-edge technology, allowing minds across the globe to contribute to and benefit from this collective intelligence, raising the benchmark of what can be achieved. It's this spirit of sharing that propels the artificial intelligence narrative forward, ensuring that neural networks remain a cornerstone of AI's thrilling expedition into the future.

Looking forward, the integration of quantum computing with neural networks tantalizes the AI domain with even more revolutionary prospects. Quantum computing's potential to process gigantic numbers of computations simultaneously could redefine the limits of what neural networks can achieve. Harnessing quantum algorithms may empower models to handle complexity and ambiguities in ways that classical computers are fundamentally unprepared for, laying the groundwork for breakthroughs we can scarcely imagine today.

As we venture into the unknown, these networks may hold the key to understanding more profound questions about intelligence itself. They challenge us to reflect on not just how we think, but also why, inviting us to ponder the very nature of consciousness and the potential for artificial life. This exploration isn't just a technical one; it's a journey that blurs the lines between science and philosophy, illuminating paths yet traversed in our quest to comprehend the universe.

Thus, "Neural Networks and Beyond" embodies more than a chapter in the saga of machine learning and deep learning. It encapsulates a story of human ingenuity and the relentless pursuit of knowledge, charting the trajectory from mimicking intelligence to collaborating with it. The confluence of these advancements with societal aspirations leads to a profound symbiosis, unveiling a future where the essence of intelligence is not only understood but endlessly expanded upon. And amidst these sweeping changes, it becomes increasingly clear: we are not merely observing the dawn of a new era; we're actively crafting it.

CHAPTER 4: NATURAL LANGUAGE PROCESSING (NLP)

Natural Language Processing, often abbreviated as NLP, is a remarkable fusion of linguistics and technology, allowing machines to understand and interact with human language as if they were conversing with us on an innate level. In this vibrant dance of syntax and semantics, NLP models strive to decipher the subtle intricacies of words, context, and intent, crafting a bridge between digital and human communication. It's the technology behind chatbots that greet you online, virtual assistants that obey your spoken wishes, and translation apps that evoke the spirit of Babel, tearing down linguistic barriers with astonishing precision. As AI continues to evolve, NLP is not only reshaping how we communicate with machines but also how we communicate with each other, turning whispers of potential into transformative realities.

Language Modelling

Language modelling stands as a fundamental pillar in the landscape of Natural Language Processing, shaping how machines comprehend and generate human language. At its core, it involves predicting the likelihood of a sequence of words, enabling a myriad of applications from autocomplete features to sophisticated text generation systems. Imagine a world where machines not only understand the syntax but grasp the myriad nuances of human speech, all driven by intricate statistical models and deep neural networks. These models, with their unparalleled ability to mimic and predict language patterns, have opened the door to intuitive interactions between humans and machines. They're the magic behind the scenes, transforming raw text data into meaningful insights and capabilities, pushing the boundaries of what we thought possible. As we stand on the precipice of AI's next leap, language models are not just tools—they're the storytellers, bridging the gap between silicon and sapiens, promising a future where machines might one day converse as naturally as our fellow humans.

From Text to Speech examines a significant and fascinating aspect of Natural Language Processing (NLP) known as language modelling. At its core, this sub-section isn't just about translating words into audible form; it's about crafting a voice that brings text to life in a way that resonates with the listener. Known as Text-to-Speech (TTS), this technology hinges on sophisticated models that predict and generate human-like speech from written input. Gone are the days when robotic voiceovers were the norm; today's TTS systems are capable of producing speech with nuanced intonation and rhythm that can mimic human expressiveness, making it increasingly challenging to distinguish from actual human conversation.

The journey from text to speech involves several intricate steps, each playing a vital role in casting written words into convincing vocal renditions. Language modelling ignites this transformation by analysing and understanding the structure and semantics of the text. The process involves interpreting the text's phonetics, prosody, and even sentiment to create a realistic auditory experience. Advanced language models like deep neural networks enable systems to adapt voice characteristics such as pitch, speed, and emphasis, allowing for a customisable and highly versatile tool.

Emulating human speech involves more than just converting text. It's about grasping the context, sentiment, and even the emotional weight behind words. Language models analyse text to ascertain these underlying nuances. Take sarcasm, for instance: a simple statement could convey an entirely different meaning when spoken with the right intonation. This complex understanding makes TTS systems adaptable across a wide spectrum of applications, from audiobooks that need to capture the nuanced expressions of characters, to assistive devices that cater to the needs of individuals with visual impairments.

Now, you might wonder how these systems achieve such a high degree of realism in speech. It's a blend of linguistic theory and data-driven approaches. Linguists provide foundational knowledge about how humans produce and perceive speech, which is invaluable for constructing phonemic architectures within TTS systems. On the other hand, vast datasets and machine learning algorithms allow for pattern recognition and prediction, driving the TTS technology to new heights. This synergy between human expertise and machine capability underscores the potential of AI technology to enrich human experiences profoundly.

Furthermore, advancements in language modelling have opened the door to personalised speech synthesis. Through the analysis of unique vocal patterns and characteristics, technology can now create bespoke voices specific to individual needs and preferences. Imagine having the voice of a loved one as your GPS navigator guiding you through traffic, or perhaps a TTS device that speaks in a tone perfect for each occasion, whether commanding authority or expressing subtle humour. Language modelling lays the groundwork for these customisable experiences, transforming our interactions with machines into more personal engagements.

But it's not just about luxury or convenience. For those with speech impairments or who require language translations in real-time, TTS offers tangible improvements in quality of life. It empowers individuals by granting access to facets of communication they may have previously found challenging, further highlighting the societal benefits of this technology. In education, TTS paves the way for more inclusive classrooms where students with reading difficulties or language barriers can engage with learning materials just as dynamically as their peers.

In terms of technical innovation, the excitement around TTS technology is palpable. Cutting-edge research continually seeks to refine voice synthesis to make it more natural and emotionally resonant. Techniques like waveforms for voice synthesis and the fusion of emotional AI are pushing the boundaries of what's possible. Researchers strive to incorporate emotional intelligence into TTS, enabling it to comprehend and react to the emotional undertones of different contexts, thus enriching the listening experience.

The implications of these developments are not limited to personal and social spheres. Industries across the board are leveraging TTS to enhance customer service interactions, streamline user interfaces, and innovate new ways to engage consumers. By integrating TTS with virtual assistants, businesses provide customers with seamless, conversational interactions that enhance satisfaction and efficiency. As AI continues to evolve, we're bound to see even more sophisticated applications emerge, each pushing the frontier further.

Despite these advancements, challenges remain in creating an entirely flawless synthesised voice. Regional accents, dialects, and the intricacies of natural human expression present ongoing hurdles. However, these challenges drive endless research and innovation, ensuring TTS will continue to captivate and surprise us in unimaginable ways. The pursuit of perfect human speech synthesis remains an evolving dance of artistry and engineering, promising an exciting future.

In conclusion, the journey 'From Text to Speech' is a testament to the phenomenal capabilities of AI within the realm of language modelling. The advancements made in this area aren't just technical achievements; they're milestones in our ongoing quest to bridge the gap between humans and machines. As we continue to explore this exciting frontier, the potential to improve and transform human-tech interactions grows exponentially, encouraging us to dream of a future where communication knows no bounds.

Applications of NLP

Natural Language Processing (NLP) is swiftly transforming the way we interact with technology, infusing a sense of human-like comprehension into digital systems. Its applications stretch across various domains, enhancing efficiency and user experience in ways that once belonged solely to the realm of science fiction. Crafting seamless interactions through chatbots and virtual assistants, NLP brings a dash of magic to customer service, handling countless queries with grace and precision. Beyond this, it powers sentiment analysis to gauge public opinion, bringing valuable insights into brand perception. Educational platforms harness its capabilities to tailor content for individual learners, creating a more personalised educational journey. Moreover, businesses are tapping into NLP to streamline operations, from automating mundane tasks to providing actionable intelligence from unstructured data. As NLP continues to evolve, its potential seems boundless, making the once unimaginable within our grasp while redefining our technological landscape.

Chatbots and Virtual Assistants are something we often encounter, but do we really think about their complexity or potential? Subtle yet profound, these AI-driven entities represent a significant leap in the application of Natural Language Processing (NLP), turning text and language-based interactions into seamless conversations or tactile responses. Let's delve into how these fascinating tools are transforming various aspects of our lives and why they are much more than mere conveniences.

Chatbots and virtual assistants are, in essence, the modern-day communicators. Gone are the days when interacting with technology felt as cold and rigid as conversing with Darwin's dodo. Today, NLP transforms cold data into warm, human-friendly exchanges. From directing users to the best routes home to providing instant customer support, chatbots and virtual assistants exemplify how language models understand context, nuance, and human emotion.

The magic lies in their ability to process and respond to the vast complexity of human language. This involves parsing sentences, recognising speech, and even identifying sentiment. It's easy to underestimate the sophistication behind a chatbot urging you to try a new café. But that little nudge represents an intricate ballet of algorithms performing a heartfelt dance of understanding and prediction.

But how do they work? In short, it's a combination of rule-based models and machine learning. While rule-based chatbots follow predefined paths, today's advanced assistants employ machine learning to comprehend richer, more natural interactions. This means instead of being limited to scripted instructions, they learn from conversations, continuously enhancing their contextual understanding. They're evolving entities, adapting, and becoming more aligned with our expectations as time unfolds.

Consider Apple's Siri or Amazon's Alexa. These aren't just voice-activated tools; they're emblematic of a paradigm shift in human-computer interaction. While accessing information through screens has been the norm, virtual assistants bring a conversational, human element to technology. It's akin to having a helpful companion who listens, processes, and acts on our requests. The accessibility granted by AI listening to our commands or inquiries in a simple, conversational voice lowers the barrier between humans and technology. This allows us to reimagine what it means to interact with digital systems.

In the business world, chatbots are revolutionising customer service. Imagine you no longer need to wait interminably for a human agent. Instead, an intelligent chatbot deals with inquiries round the clock, refining its responses over time with every interaction. This kind of efficient customer service builds stronger trust and satisfaction. From retail to banking, every sector sees a boost in efficiency and customer satisfaction through well-designed chatbots and virtual assistants.

Personal interaction is another frontier where these AI wonders dazzle. Take mental health for instance. Virtual assistants now offer companionship, providing guidance and support that once seemed far-fetched. They can't replace human therapists, but they fill a crucial gap, providing accessibility when professional help is out of reach. This non-judgemental, patient nature of bots presents an opportunity for all to access support without stigma.

Education, too, is experiencing a change. Picture a virtual assistant guiding students through complex topics, providing tailored responses to their questions. This form of personal tutor, available at any hour, enhances traditional education methods, making learning a more engaging and dynamic process.

However, it's not all seamless. The quest for flawless conversational AI is fraught with challenges. Understanding context, sense of humour, or cultural nuances can sometimes lead to unexpected or amusing errors. The key is achieving the right balance between creativity and precision, which ensures meaningful interactions. Developers constantly work to refine these interactions, striving to remove the noise from the signal, enabling smoother and more human-like engagements.

Security and privacy also pose significant challenges. When virtual assistants listen, what happens to the captured data? Ensuring secure data handling and respecting user privacy is imperative. The convergence of AI technology and human hopes brings both potential and responsibility for ethical practices.

The exciting bit is that this is just the beginning. With advancing technologies like GPT and transformer models, chatbots and virtual assistants are poised for even greater sophistication. Imagine a future where your virtual assistant anticipates your needs before you articulate them. This foresight could transform the mundane facets of our lives into seamless, intuitive experiences.

Ultimately, chatbots and virtual assistants occupy a unique space in the AI ecosystem. They're the tangible face of abstract technology, taking the high-level theories of machine learning and NLP and putting them into our everyday lives. These are not just tools — they're stepping stones to a more integrated, communicative, and ultimately understanding relationship between humans and machines. It's a partnership, one built on the foundations of Natural Language Processing, that's evolving and improving in harmony with our modern world.

CHAPTER 5: AI IN HEALTHCARE

In the intricate tapestry of modern healthcare, AI is weaving new threads of innovation that promise to redefine how we understand and approach medical science. No longer bound by traditional diagnostic limitations, AI is empowering medical professionals with tools that rival those once dreamt of only in science fiction. Think of AI-driven diagnostic systems that can sift through complex data sets in the blink of an eye, identifying subtle patterns that may escape even the most experienced human eyes. Meanwhile, in the realm of personalised medicine, AI algorithms are being fine-tuned to decode the mysteries of our genetic makeup, offering treatments tailored with unprecedented precision. These advancements not only streamline patient care but also promise to democratise healthcare, extending its reach to underserved populations. Visualise AI as a diligent assistant, tirelessly working alongside clinicians, enhancing decision-making, and fostering a new era of healthcare where prevention takes precedence over cure. As AI continues to advance, it carries the potential to transform the very essence of medical practice, ushering in an era where innovation and compassion work hand in hand for the betterment of humanity.

Diagnostic Tools

In the realm of modern healthcare, AI-driven diagnostic tools are reshaping the landscape with their remarkable ability to assist medical professionals in identifying conditions more swiftly and accurately. These tools, rooted in sophisticated machine learning algorithms, excel at analysing massive datasets, extracting patterns and anomalies imperceptible to the human eye. From retinal scans to complex pathology, AI is rapidly becoming an ally in early detection and disease management. The potential for AI to improve diagnostic accuracy not only promises to enhance patient outcomes but also to alleviate some pressure from overstretched healthcare systems. By offering insights grounded in data, these tools empower clinicians to make informed decisions, ultimately leading to tailored treatment plans. As we delve further into the nuances of this technological revolution, the story of AI in healthcare continues, fuelled by innovation and the unyielding quest for precision.

AI in Medical Imaging is reshaping the landscape of diagnostic tools in healthcare, merging the precision of machines with the art of medical diagnostics. Imagine a scenario where AI does for radiology what Google Maps did for navigation: enhancing situational awareness and optimising the path to diagnosis.

Medical imaging, with its array of complex scans and intricate images, has always been a cornerstone of diagnostic medicine. From X-rays and MRIs to CT scans, the medical community relies on these images to interpret the human body's complex symphony. Yet, even the best-trained eyes can occasionally miss a nuance. Enter artificial intelligence.

By integrating machine learning and deep learning techniques, AI offers unprecedented accuracy in image analysis. These systems are trained on millions of annotated images, learning to distinguish between healthy and pathological patterns with impressive accuracy. For instance, AI can now detect minute traces of tumours that often elude the human eye, potentially diagnosing diseases at an earlier, more treatable stage.

Consider breast cancer screenings. AI models designed for mammogram analysis have shown potential in identifying abnormalities several years before traditional methods. Such early detection could revolutionise patient outcomes, allowing for interventions that could save countless lives.

But it's not always about taking the decision out of human hands. AI excels at augmenting human capabilities. Imagine a radiologist working alongside AI, akin to a seasoned chess master paired with a lightning-fast computer, both checking each other's moves to ensure perfection.

Let's delve deeper into how this partnership could look. In developing countries where radiologists are few and far between, AI systems can act as preliminary screening tools. These systems flag images that need a human touch, ensuring that scarce resources are utilised where they are needed most. By filtering through vast amounts of data at a much faster rate than a human could, AI can effectively expand the reach of healthcare.

The progression of AI in medical imaging doesn't end there. With the rise of synthetic data—a concept where AI generates realistic medical images—training models becomes less about possessing vast datasets and more about the quality and diversity of the data inputs. Synthetic data can fill gaps in datasets, reduce bias, and provide a sort of image sandbox for new models. It allows continuous refinement and evolution in a more controlled environment, which is especially beneficial for emerging diseases like COVID-19 where initial data scarcity was a significant hurdle.

The challenges, of course, are daunting. Ethical considerations abound, particularly with regard to data privacy. The datasets used to train these systems often include sensitive patient information, raising concerns about consent and the potential misuse of personal data. Balancing technological advancement with ethical responsibility remains a critical challenge.

Further, there's the question of interpretability. Machines can sometimes arrive at conclusions based on factors that are not immediately transparent to their human counterparts.

Developing systems that offer greater insights into AI's decision-making process is crucial, ensuring that these tools complement rather than dominate clinical judgment.

On a regulatory front, countries are grappling with how to standardise AI diagnostics. Standardisation helps ensure consistency, accuracy, and fairness in how AI is applied across healthcare systems globally. Yet, achieving such harmonisation requires collaboration across borders, industries, and disciplines. The stakes couldn't be higher; these decisions set the groundwork for future medical standards.

A future where AI routinely partners with medical professionals seems inevitable. But to realise the full potential of AI in medical imaging, ongoing collaboration will be necessary among technologists, healthcare providers, and policymakers alike. By aligning goals and maintaining a patient-centric focus, this trio can overcome hurdles and lead innovations that have the power to redefine the very essence of healthcare.

The promise of AI in medical imaging is not solely technological but also deeply humanistic—it allows practitioners to focus more on patient care and less on paperwork or pixel interpretation. With more streamlined processes, medical teams can spend more time interacting with patients, providing comfort, and discussing treatment options, thereby enriching the overall healthcare experience.

For technology enthusiasts and professionals navigating this dynamic landscape, AI in medical imaging offers a vivid example of how AI's precision, when thoughtfully applied, can coexist harmoniously with the complexities of human judgment. It is an arena where the boundaries of what's possible are tested daily, painting a hopeful picture of healthcare's future, where the art of medicine and the science of AI converge not just to save time but to save lives.

Personalised Medicine

Personalised medicine represents a paradigm shift in healthcare, thanks to the power of AI. It's about tailoring medical treatment to the individual characteristics of each patient, taking into account not just their genetic makeup, but a plethora of data, including lifestyle and even social environments. This detailed and nuanced approach could drastically improve outcomes by ensuring that interventions are as precise and effective as possible. In essence, AI algorithms analyse vast amounts of data and learn from patterns that are invisible to the human eye, predicting not only the onset of diseases but also how patients might respond to different treatments. By leveraging AI, we move towards a future where healthcare is no longer based on one-size-fits-all solutions, but is as unique as the individuals seeking treatment. Imagine a world where we can predict adverse reactions before they occur, optimising treatment plans before symptoms even start. It's an inspiring leap that promises not only to extend and enhance lives but also to realise a more compassionate and efficient healthcare system.

Case Studies have long been the backbone of educational insight, providing a vivid canvas of real-world application. In the context of AI in healthcare, and more specifically, personalised medicine, these narratives become considerably more transformative. Imagine a world where medicine isn't just one-size-fits-all but is instead delicately tailored to the individual nuances of each patient. That's the power of personalised medicine driven by AI, illustrated through well-documented case studies that capture both the breadth and depth of its impact.

Let's delve into the story of a middle-aged woman named Sarah, diagnosed with a rare form of cancer. Standard treatments offered little hope, but an AI-driven platform analysed her genetic data along with lifestyle choices and medical history. The system harnessed vast databases of genomic information, drawing links between her specific mutation and successful treatment protocols found in the global pool of patient data. The result? A custom treatment plan, unique to Sarah, led to a significant regression in her tumour, opening new doors for not only her recovery but also for thousands with similar profiles.

The narrative continues with the story of children suffering from congenital heart disease. Traditional diagnostic practices often entailed lengthy processes and invasive methods. Enter an AI-based diagnostic tool, which utilises deep learning algorithms to analyse images and data from previous patients worldwide. In a hospital in London, this technology enabled paediatric cardiologists to swiftly pinpoint anomalies, reducing the timeline for diagnosis and allowing immediate bespoke intervention strategies. Families and healthcare providers alike observed a considerable improvement in treatment outcomes. The technology not only saved time but also lives.

In a radical shift from treatment to prevention, we find the collaboration of AI with wearable technology. Consider Jonathan, a 45-year-old with a history of hypertension. Extended wear of a smart device monitored his vitals in real-time, feeding data continuously into an AI system that predicted risk patterns and alarmed for preventive measures. The AI system drew correlations from vast population-level health datasets, identifying Jonathan's early warning signs ahead of a potential cardiovascular event. With timely lifestyle modifications and strategic medical interventions, Jonathan was living proof of AI's potential to prevent rather than cure.

On another front, personalised medicine backed by AI has transformed the approach to managing chronic diseases like Type 2 Diabetes. Traditional regimens usually involved a rigid medication plan, often leading to patient non-compliance due to side effects or lifestyle disruption. AI-driven systems now assist in crafting adaptive treatment plans that consider patient-reported outcomes, real-time blood sugar monitoring, and even psychosocial factors. The algorithm dynamically adjusts recommendations, much like a seasoned physician, but at a massive computational scale, resulting in improved blood sugar control and enhanced quality of life.

Venturing further afield, AI has revolutionised clinical trials, particularly in how patients are selected for experimental treatments. Matching the right patient to the right trial has historically been a logistical and ethical challenge. AI systems now automate the scan of biomedical records and genomic data, efficiently identifying candidates who are most likely to respond favorably to innovative therapies. The resultant streamlining accelerates the pace at which new treatments are validated and brought to market. This system's efficacy was particularly spotlighted during the COVID-19 pandemic when rapid testing and deployment of treatments became imperative.

Personalised medicine extends its reach into mental health care as well. Understanding and predicting mental health episodes traditionally relied on patient self-reporting and clinician interpretation. AI tools now predict episodes of mental illnesses such as depression or anxiety by analysing behavioural data from smartphones or other passive monitoring devices. Users like Emma, a university student with a history of depression, benefit from this technology. The AI flags early signs of potential episodes, offering Emma intervention strategies personalised through her digital interaction patterns before her symptoms escalate.

The broader implications of these case studies are profound. They suggest a future where healthcare is less reactive and more predictive and where each patient's journey can be uniquely sculpted by insights derived from millions of others like them. The fusion of AI and healthcare promises not only to transform how we treat disease but also how we perceive wellness. It's a world where knowledge isn't just power; it's a lifeline.

Adopting AI in personalised medicine, as illustrated by these case studies, is not without its challenges. Ethical concerns about data privacy, inclusivity in AI model training, and the digital divide in global healthcare accessibility are topics of intense discourse. Yet, these challenges are but motivators for continued innovation and discourse. The narrative of AI in healthcare is an unfolding story, and each case study adds a new dimension, highlighting not only the potentials but also the hurdles of integrating AI into personalised healthcare solutions.

These stories, rich with breakthroughs and learnings, underscore a poignant truth: AI in personalised medicine is undeniably an arena of remarkable potential. It compels us to reimagine what's possible when human expertise harmonises with machine intelligence. In doing so, it paints a future landscape where healthcare transcends standard protocol and truly embraces the individuality of each human being. It's not just about clinical outcomes; it's about championing a world where medicine evolves to meet the unique tapestry of every life it touches.

CHAPTER 6: AI IN FINANCE

In the pulsating heart of global finance, AI emerges as the quiet revolutionary, subtly reshaping the landscape with a blend of precision and predilection. It subtly orchestrates stock markets, crunching unwieldy data faster than a seasoned day trader, making algorithmic trading an art form where human intuition meets machine ingenuity. In banking, AI has become a digital detective, tirelessly trawling through transactions to sniff out fraud with unfailing accuracy, safeguarding coffers with an invisible hand. The financial realm, historically resistant to change, is finding innovation in AI-driven market predictions, which offer a glimpse into the future adorned with pixel-perfect clarity. Unlike the speculative crashes in centuries past, these predictions are tempered by AI's ability to analyse trends, bringing stability and foresight to a sometimes volatile domain. At every digital turn, AI is not only enhancing security measures but also redefining the banking experience. However, this digital metamorphosis is only in its infancy, poised on the precipice of further potential where opportunities for growth and integration beckon, quietly promising a transformation imbued with both elegance and efficiency.

Algorithmic Trading

At the intersection of finance and artificial intelligence, algorithmic trading exemplifies a profound leap forward in technology's impact on financial markets. Armed with AI and machine learning capabilities, these algorithms can process vast amounts of data at speeds unimaginable just a few decades ago, making split-second decisions that can make or break fortunes. Traders now leverage these smart systems to analyse market conditions, detect patterns, and execute high-frequency trades with a precision that humans can't match. In a realm where milliseconds matter, AI's influence reshapes the strategic landscape, optimising portfolios and managing risks with an efficiency previously unthinkable. The potential of algorithmic trading isn't just in outperforming traditional methods; it's in redefining what we see as possible, blurring the lines between human intuition and machine calculation to discover new pathways in the financial wilderness. As AI continues to evolve, the transformative power it holds for finance becomes a testament to how technology can alter paradigms, pushing the boundaries of what's achievable in markets worldwide.

Market Predictions lie at the heart of algorithmic trading—a domain where human intuition meets machine precision. The financial markets have always been a crucible of human ambition and unpredictability, but with AI stepping into the arena, we're witnessing a seismic shift. Algorithmic trading has transformed from a buzzword to the operative backbone of modern finance, where machine learning models are used to forecast market trends, making predictions about stock price movements faster and arguably more accurately than any human could achieve.

At the epicentre of these market predictions is the ability to process vast amounts of data—something that AI excels at. Financial data is generated at an astonishing rate. Every second, untold gigabytes of information flow from stock exchanges, news outlets, social media platforms, and economic reports. For human traders, parsing such a deluge of information quickly becomes overwhelming. However, an algorithm trained to recognise patterns can sift through this data jungle with ease, extracting meaningful insights that inform investment decisions.

Interestingly, these algorithms aren't just about crunching numbers. They can factor in qualitative data, such as news sentiment and market mood, further refining their predictive capabilities. By analysing how certain news headlines affect stock prices or how social media discussions can signal market anxiety or confidence, AI is beginning to touch upon something once thought to be inherently human: the understanding of nuanced context. In essence, AI is no longer just following the numbers; it's learning the language of the market.

However, the path to accurate market predictions isn't straightforward. It involves a complex mesh of models, each trained on different datasets and optimised for different objectives. At times, a single wrong input or unforeseen market event can lead to flawed predictions. Developers of these algorithms must remain vigilant, continuously refining models and factoring in new variables. The changing climate of global economies, geopolitical tensions, or unexpected financial crises can all shape the market in unforeseen ways, challenging even the most sophisticated prediction models.

Consider the adaptability of these models. They're not static; rather, they evolve with the markets. Through continuous learning processes, AI systems can adjust strategies in real-time, often without human intervention. This ability to adapt quickly to new data is instrumental, especially in environments as turbulent as financial markets. Contrary to traditional trading, where past performance might not be a reliable indicator of future success, algorithmic models can recalibrate their strategies in response to fresh data, making them more resilient to dynamic market conditions.

Yet, with great power comes great responsibility. The deployment of prediction algorithms also raises ethical and regulatory questions. The potential for market manipulation, albeit unintentional, becomes a risk with autonomous systems. If a model misinterprets a signal, it might trigger a chain reaction, exacerbating market volatility. This necessitates a robust governance framework where algorithms are regularly audited and overseen to ensure they operate within acceptable boundaries.

Furthermore, algorithms that make market predictions are often a black box. Their decision-making processes can be intricate, involving layers of neural networks, which aren't always easy to interpret or explain. This opaqueness becomes problematic, especially in scenarios where accountability is needed. Regulators and financial watchdogs demand transparency, urging AI developers to provide explanations for the predictions and decisions made by their algorithms. Efforts in developing explainable AI (XAI) are increasingly gaining traction in the finance sector, striving to bridge the gap between complex algorithms and human comprehension.

The potential of market predictions extends beyond mere forecasting. Advanced predictive models are paving the way for automated trading systems that execute trades based on signals derived from predictions. This automation can substantially increase trading efficiency, reducing latency and reacting to market changes almost instantaneously. Traders are no longer just reacting to market events - they're anticipating them, creating a proactive rather than reactive approach to trading.

Despite the remarkable progress in predictive capabilities, AI doesn't close the door on human traders—it opens new possibilities. Traders are evolving from execution roles to strategic overseers, where their expertise is complemented by AI's prowess. This partnership enhances decision-making, allowing a blend of human intuition and machine precision to coexist, each enhancing the other's strengths while mitigating weaknesses.

In conclusion, algorithmic trading driven by market predictions is redefining the financial landscape. Through the lens of AI, markets are not just a cacophony of numbers and charts but a tapestry of stories waiting to be read and understood. The potential for AI in finance is vast, and the journey is just beginning. As technology progresses, the fusion of human insight with machine intelligence will only deepen, promising a future where market predictions are as much about understanding human behaviour as they are about numbers—a world where finance and AI grow ever more entwined in a complex dance towards the future.

Fraud Detection

Fraud detection in finance has undergone a seismic shift with the advent of AI, bringing about a level of sophistication previously unimaginable. Gone are the days when financial fraudsters could easily slip through the cracks of traditional monitoring systems. Now, AI employs machine learning algorithms that tirelessly sift through vast datasets, identifying anomalous patterns and unusual behaviours that might suggest fraudulent activity. Think of it as having a tireless sentinel guarding the gates of your financial security, continually learning and adapting to new threats. This technology doesn't just react—it anticipates, providing a proactive shield against ever-evolving deceptions. As a result, financial institutions are not only tightening their security but also restoring consumer confidence by ensuring their funds and data are secure. In this dynamic landscape, AI is rewriting the rules of the game, creating a resilient financial ecosystem where trust isn't given by default but earned through unwavering vigilance.

AI in Banking Security has rapidly become an essential area of focus within the broader context of AI in finance. With the growing complexity of financial transactions and the increasing sophistication of cyber threats, banks have turned to artificial intelligence as a critical line of defence. AI's ability to detect and prevent fraudulent activities is redefining security in the banking sector, offering innovative solutions that were once considered the realm of science fiction.

In the world of financial services, security concerns are ever-present, and the stakes are high. Fraud can lead to significant financial losses for both banks and their customers, undermining trust in the entire financial system. Enter AI—a transformative force providing banks with the technology to scrutinise transactions, analyse user behaviour, and detect anomalies with unparalleled precision. By employing machine learning algorithms and neural networks, AI systems can sift through massive datasets at lightning speed, recognising patterns that might elude human analysts.

At the heart of AI's prowess in banking security lies its ability to handle vast amounts of data swiftly and accurately. Banks process billions of transactions every day, each of which presents a potential risk. Traditional rule-based systems, which rely on predefined rules and thresholds to detect fraud, often fall short due to their inability to adapt to new fraud patterns. AI, however, thrives in dynamic environments. Machine learning models can continuously learn from new data, enhancing their ability to predict and identify fraudulent activities as they evolve.

It's worth noting that AI's influence extends beyond mere detection. It also plays a crucial role in prevention. Predictive analytics powered by AI can anticipate fraudulent behaviours before they occur, allowing financial institutions to take preemptive actions. This ability to predict is invaluable given the ever-evolving nature of fraud tactics. Machine learning models can incorporate inputs from a wide array of sources, including transaction history, device information, and even social media activity, to build a holistic view of potential threats.

Collaborative filtering, a technique borrowed from the world of recommendation systems, is another method used in AI-driven fraud detection. By analysing patterns of similarity between users and transactions, banks can identify common indicators of fraud. For instance, if a user's transaction behaviour suddenly aligns more closely with known fraudulent activity patterns rather than their own historical behaviour, the system can flag this anomaly for further investigation.

AI's role in banking security extends beyond the technological sphere, also impacting regulatory compliance. Financial institutions must adhere to stringent regulations designed to protect consumers and maintain market stability. AI systems offer tools to ensure compliance by automatically monitoring and auditing transactions in real-time, identifying suspicious activities and generating reports that can be used for regulatory purposes.

Despite its significant advantages, the implementation of AI in banking security is not without its challenges. One of the primary concerns is ensuring the transparency and interpretability of AI models. For many financial institutions, it's crucial that AI systems are not black boxes—systems whose decision-making processes are opaque. Understanding how and why an AI made a particular decision is key to addressing issues of trust and accountability. Financial regulators often require explainability in AI systems to ensure adherence to legal standards and ethical guidelines.

Moreover, the ethical considerations of using AI in banking cannot be overlooked. Balancing the privacy of individuals with the need for security is a continuing challenge. As AI systems require vast amounts of data to function effectively, maintaining customer privacy while employing these systems demands a delicate implementation of data governance and security measures.

On the horizon, advancements in AI technology promise even more robust banking security solutions. Emerging areas like reinforcement learning, where systems learn optimal behaviours through trial and error in dynamic environments, could offer exciting new ways to thwart fraud. Additionally, the integration of AI with blockchain technology holds potential for creating highly secure and transparent transaction frameworks.

The journey to nascent AI maturity in banking also prompts financial institutions to rethink their workforce. As manual processes become increasingly automated, the roles of human analysts evolve. Instead of replacing humans, AI augments their capabilities, enabling them to make better-informed decisions and focus on complex problem-solving tasks that require human intuition and reasoning.

AI's transformative impact on banking security is both profound and inspiring. By merging sophisticated algorithms with the agility of machine learning, banks are better equipped to combat fraud, safeguard assets, and ensure customer trust. Looking ahead, the ongoing collaboration between AI researchers, financial experts, and regulatory bodies will be pivotal in harnessing technology's full potential, ultimately crafting a banking environment that is not only secure but also innovative and resilient against the challenges of the future.

CHAPTER 7: AUTONOMOUS VEHICLES

The road to a future dominated by autonomous vehicles isn't merely paved with technological innovation but also lined with a blend of anticipation, trepidation, and ethical quandaries. These self-driving marvels promise to redefine not just travel, but the very fabric of urban life, conjuring visions of seamless mobility and reduced congestion. Challenges abound, from the intricacies of machine perception in chaotic street environments to ensuring infallible decision-making that mirrors human instincts. However, it's not just about the tech; crafting legislation that seamlessly accommodates both innovation and safety is equally pivotal. Public sentiment sways between excitement and suspicion, highlighting the ethical tightrope of autonomy and human oversight. Despite hurdles, the inspirational drive of engineers and technologists fuels a vision where roads teem with smart, communicative vehicles, forever changing how we perceive personal and public transport.

Self-Driving Technology

Self-driving technology, a mesmerising blend of cutting-edge sensors, algorithms, and decision-making processes, is ushering in a new era of mobility. It's an extraordinary arena where machines gather and process vast amounts of data in real-time to navigate a complex world without human intervention. This technology integrates a mix of hardware such as LiDAR, cameras, and radar, and sophisticated software designed to mimic the perception and decision-making of a human driver. At its heart, machine learning models power these systems, continually refining and learning from an ever-growing dataset to enhance safety and efficiency. While we marvel at the potential of a future defined by autonomous vehicles — replete with reduced accidents, optimised traffic flow, and increased accessibility — challenges certainly remain. Yet, in this space where technology meets the road, the aspirations are as vast as the technological advancements fueling them, promising a transformative impact on both our daily lives and broader societal structures.

Challenges and Innovations In the world of autonomous vehicles, self-driving technology stands as both a marvel and a battleground of modern science. The journey to achieving fully functional self-driving cars parallels the excitement and complexity of perfecting human flight a century ago. As we dig deeper into the challenges and innovations of self-driving technology, we unmask an intricate tapestry of technological hurdles interwoven with creative solutions.

One of the foremost challenges in self-driving technology is the intricacy of perception. Autonomous vehicles rely heavily on a combination of cameras, radar, and lidar to create a detailed map of their surroundings. Each of these technologies serves a unique purpose, but integrating them into a harmonious system poses a significant challenge. Imagine a car navigating a complex urban environment, with its sensors needing to distinguish between pedestrians, cyclists, and various obstacles.

Innovations in sensor technology have been groundbreaking. Companies have made strides in developing high-resolution sensors that provide a richer stream of data for the vehicle's AI to process. This enhancement is like moving from watching a grainy black-and-white TV to experiencing a vibrant 4K resolution. Yet, despite these advancements, there remains a crucial need for refinement in sensor fusion—ensuring that data from disparate sources converges accurately and efficiently.

Another pivotal area is decision-making under uncertainty. Self-driving cars must make real-time decisions in unpredictable conditions, from changing weather and unforeseen roadworks to erratic human drivers. The decision-making algorithms need to account for countless scenarios that could unfold in a blink of an eye. It’s akin to playing the world’s most intricate game of chess, where new pieces and rules are introduced in real-time.

Advanced machine learning and AI models have transformed how these vehicles process information and make decisions.

Reinforcement learning, akin to teaching a dog new tricks, has been particularly useful. When a self-driving car encounters a new situation, it's like a student facing a problem for the first time, and the AI steadily learns optimal actions through trial and error over thousands of simulated journeys.

Perhaps one of the less discussed but equally important challenges is the question of infrastructure compatibility. Our existing roads weren't initially designed with autonomous vehicles in mind. Traffic signals, lane markings, and signage may all need to adapt to communicate more effectively with these vehicles. It's like trying to play a new video game on an outdated console. Sixty years of infrastructural norms could require a fundamental restructuring to keep pace with advancing autonomous technology.

In response, some cities have begun implementing smart infrastructure, embedding sensors and communication devices that 'talk' to self-driving cars. This innovation is still in its infancy but holds the potential to significantly enhance vehicle efficiency and safety. Think of these developments as laying down the railways for a new network, enabling faster and safer travel.

Another significant hurdle is regulatory and legal frameworks. As self-driving vehicles blur the lines between driver and machine, the law struggles to keep pace. Who's liable in the event of an accident—is it the owner, the manufacturer, or the AI itself? This question opens a Pandora's box of ethical and legal dilemmas. The need for clear policy guidelines is pressing, though shaping them is a minefield of negotiation and foresight.

Meanwhile, innovators are collaborating with regulators to create frameworks that balance rapid innovation with safety. Sandbox environments, where new technologies are tested under controlled conditions, are becoming popular. These are not unlike private laboratories where cutting-edge experiments unfold before being unleashed into the wider world.

Public adoption and trust also bear their own weight as challenges. For a technology promising to redefine personal and public transport, gaining public confidence is crucial. Incidents, even rare ones, involving self-driving cars can cause a setback to trust and adoption efforts. It's a delicate dance of showcasing reliability without overselling the technology's current limitations.

Efforts to innovate in this area revolve around transparent safety protocols and consistent communication with the public. Engaging with communities to demonstrate the technology firsthand often proves invaluable. People can be sceptical about buying into a future they only read about; seeing is believing, after all.

Furthermore, tackling the challenge of cybersecurity is non-negotiable. Autonomous vehicles are, effectively, mobile computers with high-speed connectivity. They are as vulnerable to cyber threats as any other internet-connected device. Protecting these vehicles against hacking is critical to both user safety and public confidence.

Innovative cybersecurity solutions are being developed, including automotive firewalls and intrusion detection systems that provide multi-layer protection. Much like a vigilant guard dog, these systems ensure that the vehicle's core controls remain inaccessible to malicious actors.

The propulsion toward innovation is an inspiring blend of human ingenuity and artificial capability. Large-scale simulations and advanced neural network architectures are pivotal in evolving the autonomous driving narrative. The energy and investment directed towards overcoming these challenges are a testament to a future that holds promise beyond mere transport solutions.

As we stand on the precipice of an automotive revolution, the challenges may appear as vast as the innovations are thrilling. The path is certainly not without its obstacles, but the pursuit of fully autonomous vehicles propels an era of technological renaissance that stretches beyond the boundaries of human imagination. With every incremental advancement, we inch closer to a transportation ecosystem that seamlessly integrates AI's potential into everyday life.

Legal and Ethical Considerations

The dawn of autonomous vehicles has heralded a new era where machines navigate our roads with minimal human intervention. However, as we delve deeper into this technological marvel, it becomes abundantly clear that the promise of self-driving cars is intertwined with a host of legal and ethical dilemmas. The advent of autonomous vehicles invites us to rethink liability, assess regulatory frameworks, and confront moral quandaries once relegated to the realm of science fiction.

Let's start with the legal landscape, which, much like the technology itself, is evolving rapidly. Traditionally, in vehicle-related incidents, liability is straightforward, often resting on the shoulders of the driver. However, when a machine is at the wheel, discerning responsibility becomes a complex issue. If an autonomous vehicle is involved in an accident, the blame could fall on multiple entities: the manufacturer, software developers, or even infrastructure providers. This intricate web necessitates new legal definitions and frameworks, ensuring accountability is clear and just.

Currently, laws governing autonomous vehicles vary from one jurisdiction to another, creating a patchwork that lacks consistency. For global manufacturers, this can mean navigating a legal minefield with each new market entry. While some regions have embraced the technology with open arms, others remain cautious, wary of the implications. This disparity underscores the need for international collaboration to establish universal regulations that can provide clarity and assurance to all stakeholders involved.

Beyond the mechanics of the law, ethical considerations loom large. A key challenge lies in programming ethical decision-making into these machines. Known as the trolley problem, this philosophical conundrum translates into real-world scenarios where autonomous vehicles might face momentous decisions during unavoidable accidents. Should the car prioritise the safety of its passengers over pedestrians? How does it weigh one life against another? These questions force us to confront uncomfortable truths about our values and the moral compass we wish to imbibe in our machines.

This isn't just abstract musing; these are the blueprints for humanity's future. The algorithms used in autonomous vehicles should reflect ethical standards that are universally accepted, but reaching a consensus on what those standards are can be as challenging as the technical hurdles in programming them. There's a pressing need for ethicists, technologists, and policymakers to collaborate, ensuring these complex algorithms align with societal values while being practically implementable.

Another layer to consider is privacy. Autonomous vehicles rely heavily on data to function, constantly collecting and analysing information about their environment. This includes pedestrian and traffic data, which, if mishandled, could lead to significant privacy breaches. Safeguarding this data while maintaining transparency on how it's used is pivotal to gaining public trust. The challenge here is to design systems that are both secure and compliant with privacy regulations, such as the General Data Protection Regulation (GDPR) in Europe.

Moreover, the data collected by these vehicles can be invaluable for urban planning and traffic management. However, the onus of ethical data usage rests on those who manage it. Clear guidelines on data ownership, access, and sharing are crucial. This ensures benefits derived from data leverage do not come at the cost of individual freedoms and rights, which forms the bedrock of ethical progress in technology.

Insurance is another sector poised for transformation. Traditional vehicle insurance models hinge on driver history and behaviour, yet in the realm of autonomous mobility, new paradigms are needed. Insurers must re-evaluate risk metrics, potentially shifting towards models that focus on software reliability and sensor functionality. This shift not only impacts insurance businesses but also affects users, as premiums may now reflect the technological aptitude of their vehicular systems rather than personal driving records.

The ethics of AI in autonomous vehicles also sculpts our social fabric. Consider employment; the rise of self-driving technology poses threats to many who rely on driving for their livelihood. Policymakers need to strategically think about transitions and support for affected workers. At stake is not just the future of work in the transport sector, but the very structure of economies that rely heavily on human-driven logistics and services.

Public perception can't be overlooked either. Building trust in autonomous technology is crucial. Past incidents involving self-driving cars have spotlighted both their vulnerabilities and societal readiness to accept machine judgment. Transparent communication from developers and manufacturers about the capabilities and limitations of these technologies can help bridge the trust gap. This includes being upfront about testing, failures, and improvements as part of a commitment to safety and ethical standards.

Lastly, as we march toward a future where autonomous vehicles dominate landscapes, there's a societal obligation to ensure they're accessible to all, not just a privileged few. Equity in access, ensuring that advancements in mobility technology enhance rather than exacerbate existing societal divides, is vital. Crafting policies that promote inclusivity and equal opportunity within this burgeoning sector can safeguard against potential injustices associated with technological advancement.

The journey of integrating autonomous vehicles into everyday life is more than a technological challenge—it's a moral and legal enterprise. It demands a reevaluation of not just roads and traffic laws, but the routes of human values and society's collective ethics. By confronting these issues head-on, we can steer the development of autonomous technology in a direction that's not just pioneering, but also progressive and inclusive, forever changing the course of our shared journey.

CHAPTER 8: AI IN MANUFACTURING

In the bustling world of manufacturing, AI is revolutionising the way we create and assemble products, acting as both a catalyst for efficiency and a pioneer of innovation. No longer bound by the constraints of traditional methods, smart factories powered by AI and robotics are transforming assembly lines into dynamic, responsive systems capable of adapting to real-time data. This transformation isn't just about speed; it's about precision and adaptability. With AI at the helm, predictive maintenance ensures machinery runs like clockwork, reducing downtime and cutting costs. Quality control reaches new heights as AI systems tirelessly inspect products with an unmatched attention to detail, far surpassing human capabilities. The integration of AI in manufacturing marks a new era where machines don't just execute tasks, but continuously learn and improve, driving an evolution that blends human ingenuity with technological prowess. It's not just evolution—it's revolution.

Automation and Robotics

In the bustling world of manufacturing, the dance between automation and robotics is nothing short of transformational. Picture a factory floor where robots and humans work in a seamless choreography, each movement orchestrated by AI technologies that ensure precision and efficiency. This isn't just about replacing human hands with mechanical ones; it's about enhancing possibilities, optimising processes, and driving innovations that were once fantasies. Robots, powered by AI, are now adept learners rapidly mastering new tasks and adapting to changes in real-time. They can tirelessly perform repetitive tasks while humans focus on creativity and problem-solving. The potential for smart factories highlights a future where machines not only build but also learn, predict, and innovate alongside us, pushing the boundaries of what's possible in industry and inspiring a new age of manufacturing brilliance.

Smart Factories mark a revolutionary shift in the landscape of manufacturing, one that intertwines the threads of automation, data-driven insights, and advanced robotics. As we dive deeper into the realm of AI in manufacturing, smart factories stand out as the spearheads of technological innovation, heralding a future where efficiency and precision become the new normal. At the heart of these factories is the notion of interconnected systems, often described as the Industrial Internet of Things (IIoT). The idea is simple yet transformative: integrate AI with machine-to-machine communication, advanced data analytics, and robotics to create self-optimising factories.

The beauty of smart factories lies in their ability to adapt and learn. Unlike traditional manufacturing plants, these factories can dynamically change production processes in response to real-time data. It's akin to having a living, breathing organism that senses its environment and reacts accordingly. Imagine a production line seamlessly adjusting the speed of its machines based on the demand predictions or supply chain variations. The agility and adaptability offered by this AI-infused setup mean reduced downtime and improved resource efficiency. The long-term implications are profound, promising not only lower operational costs but also enhanced product quality.

One cannot talk about smart factories without delving into the role of robotics. These aren't your run-of-the-mill assembly line robots; we're discussing machines equipped with advanced AI capabilities. They don't just perform repetitive tasks with pinpoint accuracy; they learn from each cycle, refining processes and predicting outcomes. The integration of AI in robotics within these factories facilitates predictive maintenance, a preventative approach where machinery anticipates malfunctions before they occur. This predictive prowess minimizes interruptions and extends the lifespan of costly equipment, safeguarding investments.

One captivating aspect is how these factories contribute to sustainability. In an age where environmental concerns are at the forefront, smart factories offer a reduction in waste and energy consumption. By employing AI algorithms, factories can optimise energy usage down to the minutest detail. Waste reduction strategies derive from AI's ability to identify inefficiencies in production processes, thus meticulously calibrating each step to minimise material loss. Such eco-friendly approaches not only appeal to environmentally conscious consumers but also align with global sustainability goals. Factories are now part of the solution rather than the problem—a brilliant paradigm shift.

But how does this all affect the human element in manufacturing? There's a narrative that AI and automation will render the human workforce obsolete, yet reality paints a different picture. In smart factories, humans still play a crucial role, albeit a more strategic one. With machines handling the heavy lifting and repetitive tasks, human workers can transition to roles that require creativity, problem-solving, and strategic planning. The relationship between machines and humans is collaborative; each complementing the other, forging new paradigms in workplace dynamics.

Moreover, the data-rich environment of smart factories opens up new vistas for decision-making and innovation. AI systems aggregate and analyse copious amounts of data, providing insights that were previously unimaginable. With these insights, manufacturers can innovate at a pace that meets, or even surpasses, evolving market demands. Companies can test new product ideas in a virtual setting before committing to costly production changes, thus fostering a much-needed layer of innovation without risk.

The relentless march of technology in manufacturing is not without its challenges, though. The integration of AI, robotics, and data requires robust cybersecurity measures. As factories become more connected, they are also at risk of potential cyber threats. It's essential for manufacturers to invest in secure infrastructures and anticipate vulnerabilities. While AI provides the tools to fortify these networks, it also necessitates an ongoing commitment to cybersecurity, ensuring that innovation doesn't outpace safety.

Smart factories evoke a vision of the future, where cutting-edge technologies harmonise to produce goods faster, cheaper, and with minimal environmental impact. For technology enthusiasts and industry professionals, the potentialities are virtually endless. Whether it's the controlled environment that adapts at a moment's notice or robots that operate with near-human intuition, smart factories exemplify what can be achieved when AI is fully embraced in manufacturing.

The symbiosis of AI and human ingenuity within these factories inspires a future where constraints are merely puzzles to be solved. As we stand on the threshold of this new era, smart factories embody not just a technological leap, but a profound evolution in the perception of what's possible in the world of manufacturing.

Quality Control

In the rapidly evolving landscape of manufacturing, quality control emerges as an essential arena where AI demonstrates its prowess. Harnessing the robust analytical capabilities of AI, manufacturers can now detect defects and anomalies with unprecedented precision. While human inspectors might occasionally overlook flaws, AI-driven systems tirelessly scrutinise each product or component with eagle-eyed accuracy. Using machine learning algorithms, these systems continually improve, adapting to new patterns and challenges. The result is a dramatic reduction in waste and costly recalls, all while upholding the highest standards of product excellence. As the industrial landscape transforms, AI's role in quality control not only bolsters efficiency but also empowers a culture of innovation by freeing up human resources for more creative and strategic endeavours. The convergence of human ingenuity with machine intelligence crafts a new standard in manufacturing, shifting from reactive to proactive quality assurance.

Predictive Maintenance is a fascinating intersection of AI and manufacturing, specifically within the sphere of quality control. It's not just about predicting the failure of machinery; it's about ensuring products meet high standards and reducing waste, inefficiency, and downtime. AI enhances predictive maintenance by allowing manufacturers to anticipate when equipment needs attention, thereby avoiding unexpected breakdowns. Through the application of machine learning algorithms and sensor data, predictive maintenance keeps the wheels of production turning smoothly and efficiently.

Imagine a world where factories operate with minimal interruptions. Predictive maintenance makes this a reality by using AI-driven tools to detect small anomalies in the equipment's performance before they become major issues. But it's not just about cost savings. The ripple effect extends to maintaining the quality levels of products being manufactured. When machines are running at optimal conditions, the products they create are less likely to suffer from defects, ensuring that consumers receive goods that meet their expectations.

At the heart of predictive maintenance is data—lots of it. Sensors installed on machinery collect vast amounts of information regarding operation times, heat levels, vibrations, and numerous other parameters. The big data environment of modern manufacturing processes is ripe for AI intervention. With the application of advanced algorithms, AI sifts through this data to provide insights that human operators might miss. It's like giving machines the ability to tell their own stories about their health and efficiency.

The shift from reactive to predictive maintenance is revolutionary in the manufacturing sector. Traditional maintenance schedules were based on time or usage, but with AI, maintenance can be condition-based. This means interventions happen only when needed, saving both time and resources. Over time, this change in approach enhances production quality by ensuring that all components function properly without the wear and tear that might lead to defects or failures.

Consider the impact of predictive maintenance on the automotive industry. Car manufacturing involves intricate machinery and precise assembly lines that need to run without a hitch. Predictive maintenance ensures that all these machines operate seamlessly, reducing the possibility of errors in the assembly process. This not only cuts costs arising from production halts but also improves the quality of the vehicles that roll off the assembly line. Cars require a high level of precision to be safe and reliable, and predictive maintenance plays a crucial role in achieving this.

Beyond machinery, predictive maintenance's influence on quality control extends to the human element of manufacturing, too. Technicians can allocate their skills where they're most needed, focusing on solving problems rather than constantly monitoring machines that are functioning well. It boosts morale and efficiency as workers experience less stress from unexpected crises, allowing for a more controlled and thoughtful approach to any repairs or adjustments necessary.

Moreover, predictive maintenance aids in the sustainability aspect of manufacturing, which is an increasingly vital concern in today's world. By optimizing the lifespan and efficiency of equipment, companies reduce waste and energy consumption. When machines are maintained predictively rather than reactively, they often require less energy to function correctly and have a longer operational life. This contributes significantly to conservation efforts, marking predictive maintenance as a pivotal player in a sustainable manufacturing future.

The integration of AI into predictive maintenance isn't without its challenges. The implementation requires significant initial investment and restructuring of existing processes. It can be daunting for companies entrenched in traditional maintenance frameworks to make the transition. Yet, the long-term rewards often justify the upfront costs. Companies that successfully implement AI-driven predictive maintenance find themselves ahead of competition, their production lines running smoother and their products meeting quality standards effortlessly.

Another layer of complexity involves ensuring that AI systems used for predictive maintenance are themselves reliable and free from bias. It's essential that these systems are regularly reviewed and updated. Collaboration between AI experts and industry professionals is crucial, as is continuous testing of algorithms with real-world data to ensure they're both relevant and accurate. This cooperation leads to AI tools that truly benefit the manufacturing processes without introducing new issues.

In conclusion, the potential of predictive maintenance in quality control is vast and transformative. By empowering machines with the ability to communicate their needs through data, manufacturers can preemptively address issues, maintaining seamless production flow and enhancing the quality of their output. Integration of AI-driven predictive maintenance doesn't just prepare the manufacturing industry for a more efficient future; it paves the way for operational excellence. As AI continues to evolve, so too will the precision and potential of predictive maintenance, securing its place as an indispensable part of modern industry.

CHAPTER 9: AI IN ENTERTAINMENT

In the dazzling world of entertainment, AI has taken center stage, revolutionising the way we create, experience, and engage with media. Imagine a symphony composed by a machine or a film co-directed by an algorithm—this is not just science fiction but a blossoming reality. Groundbreaking AI technologies are crafting movie scripts, generating music, and even designing video games, pushing the boundaries of creativity and inspiring new art forms that were previously unimaginable. Moreover, AI is personalising our entertainment experiences; streaming services, for instance, use complex algorithms to fine-tune recommendations, ensuring you're always hooked to the next binge-worthy show or must-hear playlist. As AI continues to transcend traditional roles, it challenges and inspires artists, filmmakers, and musicians to collaborate in novel ways, blurring the lines between human and machine creativity. In this ever-evolving landscape, AI is not only a tool but a co-creator—opening doors to endless artistic possibilities and redefining the entertainment industry as we know it.

AI-Generated Content

Dive into a world where machines create as effortlessly as humans breathe, and you'll find the enchanting realm of AI-generated content. This innovative fusion of creativity and technology is rewriting the rules of entertainment, boldly weaving stories, crafting melodies, and conjuring visual masterpieces. As artists and machines harmoniously collaborate, the boundaries of human imagination stretch, offering us art that both surprises and delights. AI not only expands artistic expression but also presents endless possibilities for customisation, letting creators and audiences engage in a dialogue of ever-evolving creativity. It's a tantalising dance between code and concept, reshaping how we experience films, music, and media. The question now isn't whether AI will transform entertainment, but to what dazzling extent.

Film and Music, as they stand at the vanguard of modern culture, have become fertile grounds for showcasing the creative potential of artificial intelligence. In vivid cinematic landscapes, AI is more than just a tool; it's a new member of the creative team, sometimes even acting as director or composer. Consider AI-generated scripts, which might not yet rival the human touch but are already offering fresh angles on storytelling. Films are no longer limited to human imagination, as AI expands the boundaries of what narratives can be.

Picture this: a film co-created by an AI that generates intricate plot structures based on audience preferences and historical data of box office hits. Studios are starting to experiment with AI algorithms that sift through mountains of data to predict a film's potential success. The technology can guide script revisions, advise on casting choices, and even suggest ideal release times. It's a blend where art meets science, and the results are fascinating, even if occasionally unpredictable.

In the realm of music, AI is orchestrating compositions that were inconceivable just a few years ago. Take, for instance, platforms that analyse thousands of hit songs to construct the 'perfect' track for commercial success. Musicians can now collaborate with AI systems that learn their style and suggest new melodies, harmonies, and even lyrics. While purists may argue about the loss of human emotion, many artists embrace AI as a co-creator, one that brings fresh insights and possibilities.

The impact of AI in music isn't just confined to creation; it's also reshaping contribution. AI can mimic iconic voices or produce entirely novel ones, giving rise to a new genre of music that blurs the line between reality and imagination. Some AI programs have been trained on Beethoven and Bach, creating symphonies that, while new, resonate with the grandeur of the masters. This ability to draw from the past while creating something new is pushing the boundaries of what we traditionally consider music.

However, this technological progress isn't without its ethical conundrums. As AI-grown music and films become increasingly indistinguishable from human-created works, questions about copyright and ownership loom ever larger. If a program writes a number one hit, who owns the rights? Is it the team behind the algorithm, the label, or the AI itself? These issues challenge the very foundations of intellectual property law and demand new frameworks that acknowledge the role of AI in the creative process.

Nevertheless, AI-generated content often sparks debates around authenticity and creativity. Can a machine truly replicate the depth of human emotion conveyed in a soulful melody or a powerful monologue? While AI offers precision and consistency, some argue it lacks the ineffable qualities that make art uniquely human. Yet, in the hands of visionary artists, these tools can enhance creativity, acting as a catalyst for new forms of self-expression.

Beyond creation, AI is also revolutionising how we consume content. Machine learning algorithms are being used to tailor playlists and film recommendations with remarkable accuracy, predicting our tastes with astonishing precision. They're capable of learning from user interactions and tweaking their suggestions in real time. Music streaming platforms and video services have seen increased engagement as a result of these personalised experiences, drawing listeners and viewers deeper into curated worlds designed just for them.

These advancements raise intriguing possibilities for the future of film and music. Imagine AI-generated films where plots evolve based on viewer reactions in real-time, or concerts where AI dynamically modifies playlists to match the audience's mood. The potential for personalisation seems limitless, offering bespoke experiences that current systems only hint at.

To theorise, if AI continues to integrate into these industries, the defining cultural products of future generations may feature content where human and machine creativity blend seamlessly. The fusion might create works that are not only technically brilliant but also emotionally resonant, striking chords that resonate with a diverse, global audience.

This cultural synthesis invites us to reconsider what we value in art and entertainment. Perhaps it is not the origin of the creation that matters most, but the connection it fosters with those who experience it. As AI assists in crafting new stories and symphonies, we may find that it is not the replacement of human talent but the enhancement of it, helping to realise visions that were previously unimaginable.

Indeed, AI in film and music isn't just about replacing traditional methods; it's about complementing them in ways that enhance narrative richness and musical vibrancy. As we embrace these technologies, we'll undoubtedly discover new horizons where human creativity and machine ingenuity meet, forging paths that both challenge and inspire.

Ultimately, AI's role in film and music is still unfolding. We're witnessing the birth of an era where creativity and technology intertwine more intimately than ever. The artistry that emerges from this synergy might one day redefine entertainment as we know it, catalysing a cultural renaissance marked by innovation, inspiration, and a renewed appreciation for the limitless scope of human-machine collaboration.

Personalised Recommendations

In the bustling world of entertainment, where options are as endless as the stars, personalised recommendations have become a cornerstone of our media consumption. Thanks to AI, platforms can create tailored experiences that speak directly to individual tastes, moods, and even whims. Imagine switching on your favourite streaming service and being greeted with a selection of films and tracks that feel handpicked just for you. This isn't serendipity; it's a symphony of algorithms working behind the scenes. By analysing viewing patterns, likes, and even the time of day you're most active, AI crafts a curated experience that feels satisfying and refreshingly intuitive. It's a harmonious blend of machine learning and human emotion, revolutionising how we engage with content. In the grand theatre of digital entertainment, AI is the master conductor, seamlessly merging data with delight. As you dive deeper into this personalised universe, you discover that these algorithms don't just know what you want—they anticipate what you'll want next, sculpting tomorrow's entertainment landscape today.

Streaming Services have made an indelible mark on the world of entertainment. As artificial intelligence (AI) continues to evolve, its integration into streaming services becomes a quintessential example of how technology shapes our viewing experiences. Personalised recommendations not only offer a seamless way to discover new content but also redefine what consumption means in the digital age. These algorithms, with their ability to learn and predict user preferences, make streaming services the omniscient guides of a viewer's journey through a sea of content.

Imagine turning on your favourite streaming service and having your perfect watchlist curated for you in an instant. This isn't magic, but the result of sophisticated AI-driven systems that have quickly become the backbone of streaming platforms. They don't just suggest films or series based on a single watch—AI delves deep, analysing behaviours, times of day you watch, pauses, rewinds, skips, and countless other data points to tailor recommendations. The result? A viewing experience that feels remarkably intuitive and personal.

To understand the impact of these recommendations, it's helpful to consider how these systems function. At the heart of AI in streaming services are machine learning models designed to adapt and evolve with each interaction. These models utilise collaborative filtering, content-based filtering, and deep learning techniques to forecast what might captivate a viewer's attention next. In doing so, they create an ecosystem where content, once hidden in the vast archives of a streaming service, surfaces at precisely the right time.

It's not all about the tech, though. The power of AI in streaming services extends to its ability to enhance viewer engagement. When users find the content they love more readily, they're likely to spend more time on the platform. This engagement is a win-win for both the service provider and the viewer. For platforms, increased viewing time translates to higher customer retention rates and better metrics for attracting advertisers and content partners. For viewers, it means richer content discovery and more meaningful interactions with what they love.

Moreover, personalised recommendations fuel a fascinating dynamic of serendipitous discovery. While some critics argue that algorithms can create echo chambers, showing users only what aligns with their past choices, others point to the unexpected delights of finding a new favourite show or movie that might have otherwise gone unnoticed. By fine-tuning the balance between familiarity and novelty, AI offers viewers a stream of discoveries, broadening cultural horizons and occasionally igniting viral phenomena.

The international reach of streaming services further amplifies AI's role in shaping global entertainment culture. Services now tailor recommendations not just by genres or actors but also by regional trends and linguistic preferences. This level of customisation ensures that audiences from different corners of the world receive content that resonates deeply with their cultural contexts. As these platforms expand, AI acts as the translator—bridging cultural gaps and fostering a more connected global audience.

As with any technological integration, there are challenges. Privacy concerns, for instance, loom large. As AI systems learn from viewer data to suggest content, the question of how that data is gathered, stored, and used remains pressing. Balancing personalisation with privacy is a tightrope that service providers must walk carefully. Regulations and transparent policies are imperative to maintain user trust and ensure ethical usage of AI in these personalised recommendations.

Furthermore, the rapid advancement of AI in streaming services nudges us toward a broader discussion about the future of storytelling. When AI can anticipate viewer reactions and preferences, it opens up new avenues for content creation. Writers and creators may soon start collaborating with AI to craft stories that are dynamically tailored to different audience segments, thereby revolutionising narrative structures and formats.

In this light, the evolution of streaming services represents a microcosm of AI's vast potential across all sectors. The technology that brings personalisation to your screen is the same that could personalise education, healthcare, and more. It's a compelling reminder that AI is an enabler of individual experiences, designed to enrich and enhance human interaction with digital environments.

Ultimately, the interaction between AI and streaming services paints a vivid picture of entertainment's future. As we continue to embrace this digital age, the symbiotic relationship between user and algorithm grows stronger, shaping experiences and bridging the gap between choice and chance. This dynamic is as much a testament to the progression of AI technology as it is a promise of what's to come. And as these systems evolve, so too will our understanding of AI's role in bringing stories, music, and films into our lives.

CHAPTER 10: AI IN EDUCATION

As education evolves in the digital age, AI emerges not just as a tool but a transformative ally, revolutionising how we learn and teach. Personalised learning, once a distant dream, is now a reality as AI tailors educational experiences to individual needs, helping students learn at their own pace. Adaptive learning systems adjust content dynamically, ensuring more engaging and effective learning journeys. Moreover, AI's role as a teaching assistant is becoming increasingly apparent, offering support through instant feedback and administrative relief for educators. In classrooms worldwide, AI is not just facilitating learning—it's reshaping it, inspiring students and educators alike to explore knowledge beyond traditional boundaries. With AI-driven innovations, education is embracing a future full of promise and greater accessibility, inviting curiosity and nurturing a lifelong love for learning.

Personalised Learning

In the realm of education, personalised learning emerges as a beacon of potential transformation, powered by artificial intelligence. Imagine a classroom where each student's learning journey is tailored to their individual needs, preferences, and pace. AI analyses data in real-time, adapting educational content to ensure no student is left behind. This dynamic form of learning empowers students to take ownership of their education while enabling teachers to understand their students better. It's not just about academic performance; personalised learning nurtures curiosity and encourages lifelong learning. As AI systems continue to evolve, the dream of a fully customised educational experience seems more attainable, promising to make learning not just a necessity but an inspiring adventure.

Adaptive Learning Systems are revolutionising the way we approach education, tailoring the learning experience to meet the needs of each individual. Imagine a classroom where the pace, style, and content of learning are dynamically adjusted to suit every student's unique strengths and weaknesses. This is no longer a distant dream but a tangible reality, thanks to the power of artificial intelligence.

The traditional one-size-fits-all model of education often leaves gaps where students can either find themselves floundering or, conversely, unchallenged and disengaged. Adaptive learning systems strive to bridge this divide, ensuring that every student receives the attention they need to thrive. By leveraging sophisticated algorithms, these systems constantly assess a learner's performance, predicting their needs, and calibrating content delivery accordingly.

At the heart of adaptive learning systems lies the ability to analyse a vast array of data points to understand how a student interacts with learning materials. These systems don't just consider what a student gets right or wrong; they delve into the time spent on each task, the strategies employed in problem-solving, and even patterns of mental fatigue. This rich tapestry of information allows for a truly personalised educational journey.

Consider, for instance, how a struggling mathematics student might benefit from an adaptive learning platform. If they continually stumble over certain algebraic concepts, the system might reroute them to explainer videos, interactive problem sets, or even suggest one-on-one virtual tutoring sessions. It's like having a bespoke tutor in the digital ether, always ready to help overcome obstacles and celebrate milestones.

But it's not all about remedying shortcomings. Adaptive learning systems also have a knack for nurturing natural talents. For students zipping through modules with ease, these platforms present advanced materials that push the boundaries of their knowledge, fostering an environment of continual growth and discovery.

The real charm of adaptive systems lies in their ability to scale. Unlike traditional tutoring, which might require significant time and resources to implement on an individual basis, AI-driven systems can be rolled out on a large scale, making personalised learning accessible to numerous students simultaneously. This democratisation of personalised education holds immense promise, not just for individual learners but also for educational institutions aiming to raise overall achievement levels.

Nevertheless, the implementation of adaptive learning systems isn't without its challenges. Data privacy remains a pressing concern. These systems rely heavily on personal data to function effectively, and safeguarding this information is paramount. Educational institutions and technology providers must work hand in hand to ensure data is handled responsibly and securely, maintaining trust and upholding ethical standards.

In addition, the integration of adaptive systems within existing educational frameworks can be a complex endeavour. Schools and universities may need to overhaul curricula and retrain educators to embrace this new mode of learning. Yet, when successfully implemented, the rewards can be immense—richer engagement, improved outcomes, and a more profound understanding of how students learn best.

The global pandemic underscored the urgent need for flexible and resilient educational technologies. As schools shuttered and remote learning became the norm, adaptive systems stepped into the breach, providing continuity and a sense of normalcy for learners worldwide. This scenario was an unplanned litmus test for adaptive learning technologies, and their performance during such a critical time has only accelerated their adoption and acceptance.

Looking forward, the trajectory of adaptive learning systems is poised for exponential growth. As machine learning models become more advanced, these platforms will evolve in their ability to provide nuanced understanding and pedagogical strategies tailored to each learner. Education as a static, linear process might become a relic of the past, supplanted by dynamic, feedback-driven learning paths that reflect the complexities of human cognition.

Educational change is often slow, but AI has propelled us into an era where adaptation and personalisation occur at an unprecedented rate. Yet, amid the buzz and innovation, we must remain anchored to the core tenets of education—promoting critical thinking, encouraging creativity, and preparing students for the myriad challenges of the future.

In this exhilarating confluence of technology and pedagogy, one thing remains clear: adaptive learning systems are not just another tool in the educational arsenal. They're a revolution, reimagining how we impart knowledge and inspire lifelong learning. As we journey further into the digital age, harnessing such technologies responsibly will shape the horizons of education for generations to come.

AI as a Teaching Assistant

Imagine a world where your teaching assistant isn't just a busy grad student but an AI with a wealth of knowledge at its fingertips. This isn't science fiction; it's the new reality in education. AI as a teaching assistant can handle administrative tasks like grading and scheduling, freeing educators to focus on what they do best—teaching and inspiring students. But it doesn't stop there. These digital aides are adept at personalising learning experiences, tailoring feedback, and even providing support to students around the clock. They offer students a unique opportunity to engage with the curriculum in ways that suit their individual learning styles, from quizzing students with targeted questions to setting challenges that stimulate critical thinking. In its role, AI doesn't replace the human touch of teachers but complements it, creating a more dynamic and inclusive educational atmosphere. It's a classroom transformation that's as profound as it is innovative.

Case Studies offer a deeply insightful view into the transforming role of AI as a teaching assistant in the educational domain. These examples bring to light the invaluable contributions AI is making in classrooms around the globe, both as an assistant to teachers and a catalyst for effective learning. As we delve into the specifics, we'll explore a variety of instances that highlight AI's integration in educational settings, from primary schools to higher education, illuminating its capabilities to inspire and innovate.

In the vibrant landscape of AI-enhanced classrooms, a standout example can be found at Georgia State University, where an AI teaching assistant named "Jill Watson" was integrated into a computer science class. Created using IBM's

Watson platform, Jill was designed to handle a high volume of questions from students enrolled in an online master's programme. Initially, students were not made aware that Jill was AI, so when they eventually discovered her true nature, it sparked conversations about trust and the perceptiveness of AI in delivering reliable academic support. In a class where rapid response time is vital, Jill Watson was hailed for saving valuable time for both students and professors by efficiently handling routine queries.

Similarly, at Carnegie Mellon University, AI is being used in an innovative project called LearnLab. LearnLab employs an AI platform to analyse and predict which teaching methods yield the best learning outcomes. By analysing patterns in large volumes of educational data generated during online courses, the system determines which approaches facilitate superior knowledge retention among students. This AI-driven analysis allows educators to refine and adapt their teaching models based on real-time data, creating a more personalised learning experience that can cater to individual learning speeds and preferences.

One cannot overlook the contribution of AI in primary and secondary education environments. In the UK, a pioneering project in a group of schools in London integrated AI to assist teachers with marking assignments and offering targeted feedback. By relieving educators from time-consuming administrative tasks, AI has enabled them to focus more on cultivating an engaging classroom environment and supporting students' emotional and academic needs. This initiative not only streamlines educational processes but also ensures students receive timely and constructive feedback, essential for their growth and motivation.

A cross-continental glance to China reveals how AI is reshaping the educational infrastructure at a systemic level. In Shanghai, SenseTime, a leading AI company, collaborates with schools to implement facial recognition technology that tracks student attendance and engagement. While this initiative raises important questions about privacy and ethics, it showcases the use of AI to automate attendance and monitor students' level of involvement in classes, allowing teachers to intervene with disengaged students more effectively. This data-driven approach has transformed traditional classroom metrics into dynamic and actionable insights, promising to enhance overall student experience.

Beyond formal education, consider the realm of online learning platforms like Coursera, edX, and Duolingo, which employ AI teaching assistants in numerous capacities. Coursera, for example, utilises AI to recommend courses to users based on their learning history and personal interests, ensuring a tailored educational journey. Duolingo's AI dynamically adjusts the difficulty of exercises and quizzes in real-time, adapting to the learner's proficiency, and creating a perfectly challenging language learning curve. These platforms exemplify how AI can transcend conventional educational boundaries, providing flexible and on-demand learning opportunities for millions worldwide.

Another compelling illustration comes from the University of Southern California, where an AI system known as Project LISTEN has been developed to assist children in learning to read. This technology converts any PC into a device that can listen to a child read aloud and provide feedback or prompts when required. By complementing traditional instructional methods, LISTEN facilitates improved reading skills through interactive engagement, all the while gathering data to better understand individual learning patterns and needs.

The success stories don't stop here. In rural Africa, an innovative AI project known as "Onecourse" is being used to teach children to read and write, even in environments lacking traditional teaching resources. In these areas, AI tablets equipped with gamified learning modules engage children with captivating multimedia experiences and interactive storytelling. By providing a scalable and cost-effective solution, AI is breaking down educational barriers, granting access to quality learning opportunities to children who would otherwise be marginalised.

It's essential, however, to address scepticism and challenges associated with AI as a teaching assistant. Concerns about bias in AI algorithms, data privacy, and the reduction of human touch in education persist. Yet, these case studies show significant promise and progress, harnessing AI as a powerful tool for administering customised and responsive education. By addressing these challenges head-on and refining AI systems, we have a real opportunity to redefine educational paradigms on a broader scale.

In wrapping up our exploration of case studies where AI functions as a teaching assistant, it's clear this technology is redefining the education landscape with bold strides. By supporting educators and engaging with students in a way that was unimaginable not too long ago, AI is reshaping how we envision the future of learning. As more schools, universities, and online platforms adopt these technologies, the success stories will only continue to grow, offering a wealth of inspiration and direction for educational institutions worldwide.

CHAPTER 11: THE ETHICS OF AI

In our journey into the digital frontier, the ethics of AI emerge as a pivotal point of reflection. Here, we face the profound questions of bias and fairness, as algorithms can inadvertently reinforce discrimination, creating disparities that technology should help to erase. It's about ensuring these digital minds respect privacy, safeguarding personal data in a world constantly hungry for information. We ponder over the relentless march of automation and its impact on individual freedom while wrestling with the moral implications of machine decision-making. As AI permeates every facet of life, striking a balance between innovation and ethical responsibility becomes not just desirable, but essential. By navigating this complex landscape with care, we aim not only to harness AI's potential but also to craft a future where technology enhances human dignity and justice, rather than undermining them.

Bias and Fairness

When it comes to artificial intelligence, bias and fairness emerge as pivotal concerns that can't be ignored. The algorithms powering AI systems, while seemingly objective, often inherit the biases present in the data they are trained on. This results in technology that can inadvertently perpetuate discrimination and inequality. It's a problem rooted not only in how AI is developed but also in how society perceives and addresses these issues. Ensuring fairness in AI is about understanding these biases and striving to create systems that can mitigate them, promoting equal opportunity and justice. It's a call to action for all involved in tech development to introspect and innovate responsibly, fostering an AI landscape that respects diversity and inclusivity. By doing so, AI doesn't just mimic society but becomes a force for positive change, redefining what fairness means in the digital age.

Addressing Discrimination in AI dives into a crucial topic under the grand umbrella of Bias and Fairness. As AI systems become more entwined with our daily lives, they increasingly influence aspects of society from job recruitment to medical diagnoses. Yet, this profound integration doesn't come without pitfalls. Among these, discrimination is a prominent concern, often crept in via the biased data used to train AI models.

To understand why discrimination in AI is such a hot-topic issue, we need to acknowledge how AI systems learn. Essentially, they are crafted to identify patterns from vast datasets. However, when these datasets reflect historical prejudices or societal inequities, the AI systems can inadvertently perpetuate these biases. Picture an AI tool used in hiring processes that trains on past recruitment data where minorities were underrepresented or unfairly evaluated. The AI risks reinforcing those same biases, effectively institutionalising discrimination within automated systems.

Addressing these biases requires multi-faceted strategies. Firstly, diverse data collection is essential. Training models on data sets that better reflect the variety and richness of human experiences can help minimise biases. But it's not just the data that needs diversification; it's the teams working on these AI systems too. Diverse teams can bring multiple perspectives to identify and address potential biases that homogeneous groups might overlook. The convergence of varied ideas and backgrounds can lead to more inclusive AI solutions.

Beyond diversification, there's a need for improved transparency. If AI developers disclose the data sources and decision matrices, it becomes easier for external experts to audit systems for bias. It's akin to opening up the AI's black box and letting the sunlight in, allowing for scrutiny and improvement. What's more, open-source collaborations can spur innovation in fairness, providing ethical guidelines as frameworks for further development efforts.

Implementing fairness-aware algorithms is another promising approach. These algorithms are designed to actively identify and correct biases within datasets. For instance, techniques like re-weighting, re-sampling, or adversarial debiasing work to make AI outputs more equitable. Re-weighting adjusts the importance of data samples, re-sampling creates balanced data, and adversarial debiasing uses adversarial networks to fight bias, ensuring that AI does not privilege one group over another.

There's also a growing ethical discourse around bias mitigation in AI. Who decides what fairness means? Different societies have varying thresholds of what they consider just and equitable, turning bias mitigation into a culturally sensitive endeavor. Just as important is the question of accountability—when AI systems go astray and foster discrimination, who's responsible? Thus, stakeholders must engage in meaningful dialogue to define clear ethics and accountability structures around AI development.

While these mechanisms are encouraging, they also illuminate a deeper, related issue: the feedback loop of AI and society. AI both shapes and is shaped by societal norms. If unchecked, the way AI reflects society might reinforce existing societal grievances. It's pivotal for concepts like fairness to evolve alongside AI, necessitating constant vigilance and iteration.

In tackling AI discrimination, education and awareness remain paramount. The more people comprehend the nuances of AI and its potential pitfalls, the more advocates we'll have for equitable solutions. Educational initiatives might range from workshops for tech professionals to public awareness campaigns that demystify AI and explain discrimination's ramifications. With understanding comes empowerment, and empowered users can demand fairer AI products.

Ultimately, addressing discrimination in AI underscores a larger quest for ethical AI—a quest as complex as the technology itself. It requires commitment, collaboration, and creativity across sectors. By fostering inclusive AI, developers not only adhere to ethical standards but also enhance user trust, paving the way for innovations that embrace all of humanity's potential. So, as we march towards an AI-driven future, it's imperative we build systems that reflect our highest ideals of fairness and justice rather than our baser prejudices.

Privacy Concerns

Privacy concerns in the age of AI are as intricate as they are unavoidable. As AI systems become more pervasive and data-hungry, the spotlight shifts onto how personal data is gathered, utilised, and secured. Imagine a world where your digital footprints are not just stored but analysed to predict your future moves, purchases, or even life choices. That's both fascinating and a tad unsettling, isn't it? While some see data sharing as a necessary trade-off for personalised services, others worry about the potential for misuse and the erosion of privacy. The balance between innovation and individual rights hangs in a delicate, constantly shifting equilibrium. We've got compelling arguments from proponents who argue AI's potential to transform sectors like healthcare and transportation, but with every new application comes the vital question of consent and transparency. How do we ensure that our personal information does not become a free-for-all, and what safeguards are essential to protect against privacy breaches? The dialogue is ongoing, and while tech pioneers race ahead, society must grapple with setting the rules of this high-stakes game.

Data Security and Consent are crucial issues when it comes to AI, especially in the realm of privacy concerns. As AI systems become more embedded in our daily lives, they gather and process vast amounts of personal data. It's this very data that powers their functionalities, from predicting consumer behaviour to diagnosing medical conditions. However, with great data comes great responsibility.

When we talk about data security in AI, it's not just about keeping the data safe from malicious actors. It's also about ensuring that such data isn't misused by those who collect it. Imagine a world where an AI can predict your every move, not because it's inherently brilliant, but because it has access to every single detail of your digital life. Feels a bit invasive, doesn't it?

The first step to ensuring data security is recognising the nature and volume of data AI relies on. We're talking about everything from your browsing history to your shopping habits, even your voice patterns and facial features. This array of information is like a treasure trove for AI models, leading to improved personalisation but also raising significant privacy concerns. The balance between benefiting from AI's capabilities and safeguarding personal privacy is delicate and requires constant vigilance.

Consent plays a pivotal role in determining how data should be handled. It's not just about thrusting a lengthy terms and conditions document at users, expecting them to blindly agree. Real, informed consent means users genuinely understanding what they're signing up for. They should know how their data will be used, who will have access to it, and for how long it'll be stored. Easier said than done, right?

To foster a culture of informed consent, organisations need to rethink their data policies. They can start by simplifying their user agreements and providing clear, concise explanations of data practices. This transparency boosts trust, and in a time when data breaches make headlines, trust is a commodity that's becoming increasingly valuable.

It's also worth noting that consent in the digital age isn't a one-time checkbox. Our needs and preferences evolve, and so should our permissions. Users should have the autonomy to refine what data they share over time and have easy access to their own data. Organisations that facilitate this empowerment will likely find themselves at an advantage, both ethically and commercially.

Another layer to this discussion is the technological solutions aimed at enhancing data security. Encryption techniques have become more sophisticated, allowing data to be securely transferred and stored. Moreover, concepts like federated learning show promise by allowing AI models to learn from data without it ever leaving the user's device. Such innovations could bridge the gap between privacy and progress, though they aren't widespread just yet.

Furthermore, regulatory frameworks are stepping up to the challenge. From the European General Data Protection Regulation (GDPR) to the California Consumer Privacy Act (CCPA), various legislations are being put in place to protect consumer data. These laws are setting standards, ensuring that organisations adhere to stringent security measures and respect user consent. However, the landscape of data protection is ever-evolving, and these regulations must constantly adapt to keep up with technological advancements.

Despite these strides, challenges remain. The rapid pace of AI development often outstrips the regulation processes, leaving gaps in protection. There's also the global mismatch, where regulations can vary dramatically from one jurisdiction to another. This poses a problem for companies operating internationally, as they must navigate a maze of differing data protection requirements.

Then there's the question of accountability. Who's responsible when an AI system misuses data? Is it the developer, the company employing the AI, or the end-user? These grey areas make legal and ethical discussions around AI particularly complex. Establishing clear accountability is crucial for the future of AI's role in society.

As we wade deeper into this digital era, the societal implications of AI-driven data practices can't be understated. It's a landscape that is both awe-inspiring and fraught with potential pitfalls. Educating users about their rights and the importance of consent can shift the power dynamic, promoting a more equitable interaction between humans and machines. Technology enthusiasts and professionals alike should advocate for improved standards and responsible innovation to ensure that AI serves society ethically and effectively.

Data security isn't just a technical issue; it's a societal one. By prioritising consent and transparency, and by leveraging technological innovations, we can build a foundational trust between AI systems and users. This trust will serve as the bedrock for future advancements, ensuring that the development of AI not only respects individual privacy but thrives responsibly in the process. After all, the key to a sustainable future in AI lies in our collective ability to manage data with the respect and caution it deserves.

CHAPTER 12: AI AND EMPLOYMENT

The impact of AI on employment is a captivating tapestry woven with threads of challenge and opportunity. As machines become more adept at performing tasks traditionally done by humans, there's no denying the wave of job displacement looming on the horizon. Yet, the automation paradox reveals a curious twist—while some roles may fade, AI ushers in a plethora of new opportunities that redefine the labour landscape. This transition isn't just about survival; it's about evolution and adaptation. We witness an expansion in roles that require a symbiosis of human intuition and machine efficiency. Fields previously unimaginable are sprouting, fostering an environment where creativity and technical acumen thrive. As technology enthusiasts and professionals ponder the future, one thing's clear: adaptability will be key, and those prepared to pivot will find themselves at the forefront of the new job market valleys yet to be explored.

Job Displacement

As artificial intelligence continues to shape the landscape of employment, the topic of job displacement has emerged as a lightning rod for both concern and contemplation. While we witness AI systems executing tasks with unprecedented precision and at breakneck speed, the question that often arises is: what happens to the human workers who are outpaced by machines? Economists and technologists have long debated whether these disruptions will lead to a widespread loss of employment or a transformation that creates new roles we haven't even imagined yet. The repetitive, mundane tasks, once the staple of manufacturing floors and call centres, are rapidly being automated, so it's easy to see why anxiety looms large. Yet, history offers a hopeful lens: past technological revolutions, like the Industrial Revolution and the advent of the Internet, eventually led to greater job creation. As we stand on the brink of this AI-driven transformation, it's crucial we consider how to reskill and reposition the workforce to better align with a future where human ingenuity and AI collaboration can thrive together.

The Automation Paradox is a fascinating phenomenon and a key piece in the puzzle of "AI and Employment". It tackles the irony in automation's role in both creating economic displacement and opening doors to new possibilities. As we continue to see machines taking over tasks traditionally performed by humans, it's easy to focus solely on job losses. But if you peel back the layers, a more nuanced picture emerges. While technology does replace certain job roles, it simultaneously creates opportunities, often in ways we can't immediately predict.

At the heart of the Automation Paradox lies a fundamental tension: the transitionary pain of job displacement is often counterbalanced by the creation of new roles and sectors owing to technological advancements. Historical trends illuminate this back-and-forth pattern. Consider the industrial revolution. At first glance, the advent of machinery spelled doom for countless artisanal craftsmen, yet it eventually led to a surge in different types of employment opportunities, ranging from machine operation to industrial management and beyond. AI, as a force similar in magnitude, is following this pattern in modern times.

Automation anxiety isn't unfounded, though. With AI rapidly evolving, many jobs are becoming obsolete more quickly than some workers can adapt. Take, for instance, the roles in data entry or basic customer service. These are increasingly being handed over to sophisticated algorithms and chatbots that can perform the tasks more efficiently, and often with greater accuracy. The speed at which these transitions occur sometimes leaves workers in the lurch, feeling undervalued and underprepared. For sectors struggling to keep up, the results can be quite disruptive.

Nonetheless, while the immediate effects might feel daunting, there is a silver lining. New industries often spring forth from the embers of the old. AI development itself has revitalised sectors like data analysis, cybersecurity, and even ethics, offering tailored roles for people with the skills to manage, interpret, and apply AI in ethical ways. Moreover, automation doesn't merely nix jobs; it also enhances them. Drivers may be in shorter supply with self-driving vehicles on the horizon, but demand for fleet managers and maintenance experts is on the rise.

The Automation Paradox also highlights how businesses can adapt to utilise AI while supporting their workforce. Businesses that successfully harness automation often boost productivity, allowing them to expand and hire more staff to meet increased demand. For instance, AI-equipped warehouses may replace some manual tasks with robotics but may also increase the need for skilled personnel to oversee and innovate these operations. Moreover, enhanced AI-powered tools can refine human skills, much like how computer software elevates an architect's ability to visualise and design intricate structures.

In focusing on job displacement alone, critics sometimes overlook the necessity of re-skilling and up-skilling. These processes are pivotal in allowing displaced workers to transition into newly created roles. Educational programmes, online courses, and vocational training need to adapt and expand, ensuring the workforce can meet the changing demands. With proactive measures, AI can become a catalyst for professional growth rather than an inhibitor. Governments and institutions must therefore play an integral role in facilitating these educational initiatives.

Yet the Automation Paradox isn't all about economics and pragmatics. There's a human side too. As AI takes on more drudgerous tasks, it liberates individuals to engage in creative, strategic, and relational work that many find far more fulfilling. In fact, human-centric skills such as problem-solving, empathy, and creativity remain largely out of AI's reach, at least as of now. This means that talents oriented around these skills could grow in demand, offering a more enriching work experience for many.

Understanding and navigating the Automation Paradox requires a mindset shift—from seeing AI purely as a threat to viewing it as a transformative tool that, with the right guidance, can enhance our capabilities. Indeed, reflecting on the bigger picture, one realises that AI provides a chance to rebalance what's valued in work. It challenges us to reconsider what kind of world we want to shape and, in turn, what work will look like.

The key to mastering this paradox lies in steering AI towards complementing human work rather than supplanting it altogether. Companies must act not just as adopters of AI but as creators of ecosystems where human and machine capabilities aren't at odds but are integrated in ways that amplify productivity and satisfaction. Policymakers, educators, and innovators should collaborate to create an employment landscape that welcomes change while valuing heritage skills, constructing a harmonic future where technology and humanity prosper side by side.

Admittedly, there's no one-size-fits-all solution to managing job displacement through automation. Each industry has its own nuances and challenges. However, what binds them is the potential to innovate continually. With every wave of disruption, new islands of opportunity emerge. Ultimately, how we respond to these shifts defines the shape of our economies and societies for generations to come.

New Opportunities

In the evolving landscape of AI and employment, new opportunities are sprouting like never before, reshaping the professional milieu in ways that ignite both curiosity and hope. While there are fears of job displacement, there's an equally vibrant story unfolding where AI acts as a catalyst for novel roles and industries. From AI-specialised careers that involve designing intelligent algorithms to roles focused on ethical AI deployment, the horizon is vibrant and vast. Sectors like healthcare, finance, and creative industries are witnessing the emergence of previously unimagined positions - think AI ethicists, machine learning trainers, or even AI-assisted creativity gurus. These opportunities aren't confined to tech-savvy circles; they reach well into diverse fields requiring empathy, ethics, and human insight, showcasing the symbiotic potential when human ingenuity meets machine efficiency. The dawn of AI is not just an era of automation, but a reimagining of work itself, where human creativity and AI innovation dance together, forging pathways to prosperity and progress. As this chapter unfolds, it's essential to grasp that these opportunities are not reserved for the few; they're a call for diverse talents to engage and redefine the labour landscape with AI as a powerful partner.

Future Job Market It's the dawn of an era where AI is not just a tool but an integral part of employment ecosystems, creating new roles at a pace we couldn't have imagined even a decade ago. We've all heard the looming concerns about job displacement due to automation, but less talked about are the fresh opportunities sprouting up in the wake of these technological advancements. Picture a world where jobs that were once only a figment of imagination are becoming realities—roles like AI ethics officers, robotics maintenance specialists, and data detectives are carving out spaces in this evolving job market.

The future job market is like a complex tapestry being woven with threads of technology, human creativity, and adaptability. It's not just about survival; it's about thriving in a landscape that's constantly morphing. Consider AI trainers who are responsible for teaching machines to understand human emotion, or digital artists leveraging AI to create stunning, intricate digital art. The lines between human creativity and AI's logical capabilities will blur, opening doors to professions focused on collaboration rather than competition.

But what skills will this new job market demand? There's no one-size-fits-all answer, yet certain patterns are emerging. Tech skills, particularly around programming and data analysis, will be critical, but so will soft skills like emotional intelligence and problem-solving. Jobs in the future won't just be about what you know but how you think and how agile you are at applying what you learn to new, unfamiliar problems.

Moreover, the growing reliance on AI technology means there will be an increasing need for roles that focus on the ethical implications of AI implementation. Ethics officers or consultants will be crucial in ensuring that AI's integration into daily operations is done responsibly. We're looking at a future where understanding the ethical landscape will be just as critical as comprehending the technical one.

While it might seem that the tech industry will primarily drive future employment opportunities, AI will seep into the fabric of every sector. Within healthcare, we can anticipate burgeoning opportunities for AI consultants who'll work alongside physicians to interpret data from AI diagnostic tools. In finance, AI compliance officers may become essential, ensuring that algorithms operate within regulatory standards while balancing financial innovation.

Yet, with all this change brewing, education systems must pivot to prepare the workforce of tomorrow. Curriculums need urgent restructuring to integrate AI literacy and interdisciplinary learning. We'll see increased partnerships between educational institutions and industries, forming pipelines that produce candidates ready to tackle these emerging roles head-on.

It's a dynamic shift, akin to the transition witnessed during the Industrial Revolution, but on a scale that transcends geographical and sectoral boundaries. The job market's transformation due to AI is not confined to any one industry or profession but is rather a network of interconnected opportunities that redefine the essence of work itself.

The seamless integration of AI into everyday work isn't just a challenge but a remarkable opportunity to enhance human creativity, efficiency, and productivity. Imagine the possibilities when AI takes away mundane tasks, freeing creative minds to focus on innovation and strategy. The impact of human-AI collaboration might amplify human potential rather than undermine it.

Of course, this is not without its hurdles. Transition periods are often fraught with growing pains, and the future job market will be no exception. Reskilling will be imperative, not just to keep pace with technological advances but to ensure inclusivity in this burgeoning world. Policies will have to adapt swiftly, framing guidelines that protect both employees and employers in this uncharted territory.

Despite potential challenges, the trajectory is promising. There's an inevitable reshaping of work-life paradigms—geographic constraints will lessen due to remote working technologies powered by AI, and work culture will lean heavily towards flexibility and continuous learning.

As we stand on the precipice of this transformation, it seems the future job market will be less about competing for jobs among humans and more about pioneering alongside AI. The spark lies not in seeing AI as a threat but as a partner in innovation. It's an inspiring time to reimagine what's possible, to explore paths previously untravelled and create a workforce that leverages both human ingenuity and AI advancements.

Indeed, the future job market beckons with promises that bring together the best attributes of humanity with the expansive possibilities AI offers. It's an exciting vista filled with potential for those courageous enough to embrace the change. Let's imagine what could happen when passion, purpose, and AI combine to shape the future of work.

CHAPTER 13: AI GOVERNANCE AND POLICY

As artificial intelligence continues its meteoric rise, the task of governing such a rapidly evolving technology becomes a daunting yet crucial challenge. The establishment of regulatory frameworks is vital to ensure AI's growth aligns with societal values and international standards. Unlike traditional technologies, AI's global reach and transformative power demand co-ordinated policy efforts across borders. This chapter explores the complex interplay between innovation and regulation, weighing the importance of encouraging technological advances while safeguarding against unintended consequences. Crafting public policy for AI requires a delicate balance, threading the needle between stifling innovation and ensuring fair, ethical use. As we navigate this new frontier, we also see a mosaic of global perspectives emerging, each contributing unique insights into the rules and frameworks that will shape AI's future. The journey to responsible AI governance is ongoing, involving stakeholders from all walks of life - making dialogue and collaboration more critical now than ever.

Regulatory Frameworks

In the vast expanse of AI's potential, regulatory frameworks emerge as the essential guardrails guiding its ethical and effective deployment. As we impatiently embrace change, the need for well-structured regulations becomes glaringly crucial to safeguard personal liberties, ensure societal norms are respected, and maintain fairness across different sectors. Imagine a landscape where AI technologies flourish unchecked; the implications could range from privacy infringements to deep-seated biases in decision-making processes. Balancing innovation and regulation isn't merely a bureaucratic exercise; it holds the promise of harmonizing technological advancement with human values. Across the globe, nations are grappling with the challenge of implementing scalable and adaptive regulatory structures that respect cultural nuances while aiming for a unified approach. Crafting robust policies that transcend borders might just be the key to leveraging AI's transformative power without compromising on the rights and principles we hold dear. The delicate art of regulation, therefore, doesn't stifle creativity; it inspires an ethical evolution in AI development, guaranteeing that technology remains our ally rather than a rogue entity.

Global Perspectives When discussing *AI Governance and Policy*, the idea of *Regulatory Frameworks* isn't confined to a single national perspective. Instead, it's a vivid tapestry woven from the threads of diverse global experiences and philosophies. Each nation approaches artificial intelligence with its distinct socio-economic context, which colours how its regulations evolve. The complexity of laying down universal policies is akin to crafting a symphony with instruments from around the world, each playing to its own rhythm yet somehow contributing to a harmonious whole.

Some countries, it seems, sprint ahead, setting new precedents and testing the boundaries of regulatory possibilities. The European Union has been a trailblazer in this regard, unabashedly enacting comprehensive regulations through the General Data Protection Regulation (GDPR). The GDPR doesn't merely focus on AI, but its implications touch every aspect of digital data, thereby indirectly influencing AI governance. Meanwhile, the EU has also been working on the AI Act, a pioneering attempt to regulate AI technologies based on risk. These initiatives underscore the EU's commitment to human-centric policy-making, yet they also raise concerns about innovation stifling and bureaucratic sluggishness.

Across the pond, the United States takes a markedly different route, priding itself on fostering innovation with a relatively laissez-faire approach. Instead of stringent and immediate regulation, the focus lies more on ethical guidelines and voluntary frameworks. This stems from a broader cultural attitude valuing entrepreneurial freedom and market-driven solutions. However, this approach is not without its criticisms, particularly concerning privacy and data ethics, as companies often prioritise rapid development over comprehensive consumer protection. Balancing these needs is a high-wire act, with regulations whispering in one ear and innovation loudly beckoning in the other.

In Asia, the approach to AI policy is as varied as the continent itself. China, for instance, has positioned itself as a global leader in AI, driven by centralised government mandates and substantial investments in AI development. The Chinese government has been proactive in formulating AI plans that tightly align with its economic and social objectives, placing a strong emphasis on AI as a national priority. However, China's regulatory framework raises questions about surveillance, transparency, and the potential for AI to encroach on privacy and individual freedoms. This model, while fostering rapid advancement, ignites debates on the trade-offs between progress and civil liberties.

On the other end of the spectrum, Japan integrates its AI policy-making with its cultural ethos of harmony and human augmentation. Japanese frameworks often focus on AI as a partner to society, reflecting the nation's ambition to use AI to address societal issues such as an ageing population. Here, AI isn't just a tool but a vital cog in the machinery of societal wellbeing. Similarly, South Korea places significant emphasis on AI as a driver of economic growth, yet it takes a more cautious approach to regulation, prioritising ethical standards and guidelines to ensure that the technology respects human rights.

These differences in regulatory approaches aren't just academic—they impact the global AI landscape significantly.

Nations' strategies filter into international trade and competitive dynamics, affecting everything from tech innovation hubs to talent migration. Moreover, they're a subtle reflection of national identity, manifesting how different societies view not just technology, but their future in a world increasingly influenced by artificial intelligence.

International organisations, such as the United Nations and its affiliated bodies, also play a crucial role in advising and supporting countries to formulate effective AI policies. By facilitating dialogue among nations, these organisations aim to harmonise the global regulatory landscape. The ultimate goal is a balanced approach that maximises AI's benefits while minimising its risks. Yet, achieving global consensus is an arduous task, akin to herding cats in a technological ecosystem known for its rapid evolution and complexity.

Multilateral frameworks, like those proposed in the UNESCO recommendations on the ethics of artificial intelligence, strive to establish common ethical grounds and standards. These efforts often highlight the tension between sovereignty and the need for a unified approach to global challenges like AI. The establishment of common ethical norms can serve as guiding principles, yet their enforcement and acceptance remain stubbornly diverse. It's clear that ethics, much like regulation, isn't a one-size-fits-all solution.

Global perspectives on AI governance are continually shifting, influenced by technological breakthroughs, geopolitical shifts, and societal changes. There exists a feedback loop, where regulatory frameworks not only shape AI's evolution but are also transformed by it. As innovations emerge, they redefine the boundaries and capacities of regulatory strategies, calling for adaptive governance models that can respond to unforeseen challenges.

In crafting these frameworks, nations must contend with ethical dilemmas, privacy concerns, and the unpredictability of future AI developments. The challenge is not simply deciding what AI is or isn't permissible, but ensuring that these technologies align with universal human values. This ongoing global conversation reflects our collective attempt to navigate a rapidly changing landscape. AI isn't just a technological challenge—it's a human one, engaging the best and worst of what it means to be a part of an interconnected global community.

Understanding and appreciating these diverse perspectives becomes vital for anyone engaged in the discourse on AI. As technology enthusiasts, professionals, or curious minds learn more about the possible futures AI can carve out, they should also grasp the rich tapestry of regulatory philosophies from around the globe. These frameworks, whether regulatory, ethical, or philosophical, compose a crucial part of the dialogue that will determine AI's role in our shared future.

Public Policy and AI

The intersection of public policy and artificial intelligence is as much about imagination as it is about regulation. Consider it a complex tapestry where not only technologists and lawmakers but also philosophers and ethicists weave threads of influence.

At its essence, public policy aims to guide the development and deployment of AI technologies in ways that maximize public good while minimizing harm. Getting this balance right, though, is no trivial task. The challenge lies in drafting policies that are both robust enough to tackle risks and flexible enough to encourage innovation.

In many ways, public policy in the field of AI is still in its infancy. Some governments have moved swiftly, seeing the transformative potential of AI as both a competitive advantage and a societal necessity. Countries like Estonia and Singapore have emerged as unexpected leaders, rolling out comprehensive AI strategies that not only focus on regulation but also on seeding the next generation of AI workforce. Conversely, larger nations such as the United States and China are embroiled in an AI arms race, making the development of public policy a strategic imperative.

But what does effective public policy look like when it comes to AI? At a foundational level, it involves creating a legal framework that addresses key issues like transparency, accountability, and fairness. These frameworks often require cross-disciplinary expertise, as AI intersects with so many facets of society—from economics to ethics, and education to employment. Unlike the faster-paced world of tech innovation, policy development can be seen as a slow-moving behemoth. However, drawing parallels from established sectors like pharmaceuticals and finance can offer insights into creating effective governance models.

One element that's often debated is the need for international cooperation in AI governance. AI's borderless nature makes unilateral policy action insufficient and potentially harmful. Multinational collaborations can offer a broader perspective, sharing best practices and aligning objectives. This kind of cooperation might result in something akin to the 'Paris Agreement' for AI—a global commitment to ethical AI development. However, given the diverse interests and priorities of countries, such an agreement is easier envisioned than achieved.

The landscape of public policy is also shaped by the concerns of the public it aims to protect. Public perception of AI largely influences policy decisions. When people are fearful of losing jobs or worried about privacy invasion, policymakers tend to react more cautiously. Conversely, in societies where the potential benefits of AI—such as better healthcare or more efficient services—are well understood, policies tend to be more progressive. Hence, education plays a crucial role in shaping public opinion and, by extension, public policy.

We must not overlook the role of industry stakeholders in shaping AI policy. The tech giants, for example, wield considerable influence, sometimes even more than the governments themselves. Industry partnerships with policymakers can lead to well-rounded policies that consider commercial realities alongside societal needs. However, these collaborations need transparency to avoid conflicts of interest and ensure they serve the public good rather than entrenched corporate interests.

A fascinating aspect of public policy and AI is the debate around ethical AI. Policymakers are increasingly turning to philosophical reflections on what constitutes ethical behaviour in machines. Should AI systems always prioritise human life? How do we decide whose values these systems should reflect? Questions like these are not purely academic; they have real-world implications that public policy needs to address. Establishing ethical guidelines is foundational to ensuring that AI systems align with shared human values.

Another layer of complexity is introduced when considering AI's impact on global inequality. Advanced AI can widen the gap between rich and poor, both within and between countries. Wealthy nations could lead in AI innovation, further cementing their economic dominance, while less affluent nations lag behind. Public policy can mitigate this risk by fostering global research collaborations and equitable access to AI technologies and their benefits. Initiatives like open-source AI platforms and international academic exchange programs can play pivotal roles in levelling the playing field.

As much as AI promises a new world, it also challenges existing structures and governance models. Redefining legal and ethical norms isn't just inevitable; it's necessary. Inaction or poorly designed policies could lead to unintended consequences, ranging from biased algorithms to disruptive societal changes. Policymakers must therefore remain vigilant, adaptable, and proactive in their approach to governing AI.

The importance of pilot programs and iterative policy development cannot be overstated. Policymakers might consider experimenting with local AI governance before scaling solutions to a national or international level. Such pilot initiatives offer valuable insights, enabling adaptive policies that evolve with technological and societal dynamics. Think of them as living documents—a theoretical framework that remains agile and sensitive to real-world experiences.

Public engagement and consultation also form an essential part of the policy-making process. By involving diverse stakeholders—ranging from academic institutions to civil society and industry leaders—policymakers can create inclusive strategies that reflect a broader spectrum of interests and concerns. Open dialogues and inclusive processes contribute to creating policies that are not only more robust but also more widely accepted.

So, where do we go from here? Public policy is about more than just rules and regulations; it's about envisioning a future where AI augments, rather than dictates, human lives. In addressing the multifaceted issues of AI, from ethics to equality, public policy aims to be a guiding light, ensuring AI enhances human experience rather than diminishes it. As AI technologies speed ahead, it's crucial for our policy frameworks to keep pace, acting as the guardrails that guide us toward a future filled with promise but grounded in collective wisdom and shared values.

CHAPTER 14: AI AND SOCIETY

The intersection of AI and society is a nuanced dance, teeming with possibilities and challenges as artificial intelligence weaves itself into the fabric of everyday life. As algorithms evolve, so too does their impact on social structures, influencing everything from how we communicate to how communities form and function. The transformative power of AI in social media, for instance, reshapes how information is spread and consumed, sparking dialogues on authenticity and echo chambers. Public perception dances to a rhythm dictated by both promising advancements and cautious skepticism, as surveys and polls highlight fears alongside optimism. This digital evolution brings with it subtle shifts in how we perceive societal norms, pushing the boundaries of privacy and reshaping our cultural landscapes. As AI continues to permeate society, it's crucial to remain mindful of its implications, fostering an informed public that actively participates in the dialogue about how best to integrate these technologies in ways that enhance our shared human experience.

Social Implications

As artificial intelligence increasingly integrates into daily life, its social implications ripple through the core of society, prompting profound transformation. On one hand, AI holds the potential to democratise information and services, bridging gaps in education, healthcare, and personalised content delivery. It empowers individuals with unprecedented tools to solve complex problems, catalysing innovation at every level. Yet, this technological surge also casts shadows of uncertainty, with concerns over job displacement, privacy, and digital dependence. Society must navigate these choppy waters with care, balancing technological advancement with ethical foresight. Striking this equilibrium requires dialogue that's not just open but inclusive, ensuring that AI serves as a force for collective good rather than a driver of division. As we stand at this intersection of technology and humanity, the choices we make will sculpt the societal landscape for generations to come.

AI in Social Media occupies a peculiar position in the landscape of AI and society. This marriage between advanced algorithms and public discourse hasn't just influenced our daily interactions, but it's also begun to reshape how we understand societal norms and individual behaviours. As digital communication channels evolve, AI infuses its capabilities into the fabric of social platforms, subtly but decisively influencing both the micro-interactions of personal connections and the macro-dynamics of societal change. The social media landscape is a bustling metropolis of information, and AI serves as its guide, though not without controversy, steering conversations and predicting your next click.

Social media platforms employ AI to tailor content to individual users, making each feed an echo chamber tailored to individual preferences and biases. AI's algorithms learn from each user interaction – every like, share, and comment refines its understanding of what keeps individuals engaged. This personalisation is a double-edged sword; it keeps users tethered to their devices, informed and entertained, but it can also form feedback loops that reinforce existing beliefs and biases. Thus, the social implications extend far and wide, questioning the very foundation of objective information dissemination in a democracy-driven society.

In the realm of content moderation, AI plays a vital but complex role. Given the massive influx of posts, tweets, and videos every second, human moderation is impractical. AI steps in, identifying and flagging potential breaches in the community guidelines swiftly. Yet, the task is undoubtedly knotty; algorithms often lack the nuance needed to discern context like irony, satire, or the cultural variances that might frame particular phrases as harmless in one context but offensive in another. Missteps in moderation can lead to significant social unrest, especially when perceived bias in content moderation policies comes to light, often triggering accusations of censorship.

AI also carries the burden of ensuring safety within social media. Tools powered by machine learning identify and mitigate cyberbullying, trolling, and hate speech, creating a potentially safer environment for users. Though not foolproof, their application has proven to be a step forward in dealing with online harassment. However, false positives and negatives remain a drawback, sparking debates on AI's capability to handle such sensitive issues without human oversight. Notably, reliance on automation for these tasks leads to greater scrutiny regarding transparency in algorithmic decision-making processes.

There's a darker underside to AI in social media, one that broaches the subject of misinformation. Algorithms optimised for engagement can inadvertently prioritise sensationalist or false content purely because it generates clicks. In extreme cases, AI has been implicated in spreading misinformation during critical moments, like elections or public health crises, amplifying falsehood at a bewildering pace. The resulting misinformation can erode public trust and have dire consequences on public perception and behaviour.

On a brighter note, AI aids in enhancing accessibility on social media, offering features like automatic captioning, descriptive image texts, and translation tools. These tools break down barriers, allowing users with disabilities to participate more fully in the digital community. AI's language processing capabilities mean that real-time translation is more feasible, closing linguistic divides and fostering more inclusive conversations on a global scale.

Social media's influence extends into the realm of consumer behaviour, where AI-driven analysis of user data translates into personalised advertisements. This highly targeted marketing can produce efficient sales funnels but presents its own ethical quandaries. Privacy concerns surface here, as users often find themselves caught in the labyrinthine consent mechanisms that safeguard their data. The ethical debate centres on informed consent and data ownership, prompting wider societal introspection on the acceptable use of individual data.

Moreover, AI's ability to generate and curate content challenges notions of authenticity. Bots masquerading as human users and deepfake technology raise the stakes in discerning reality within social media. Deep learning models create content that mimics human creativity, sometimes fooling users into believing fabricated stories or videos. This presents social media platforms with an existential challenge where ensuring content credibility becomes intertwined with maintaining user trust.

As we continue to integrate AI into social media, new social norms are emerging. The sense of global community formed through digital interaction is juxtaposed with a fragmentation of public discourse. The implications are profound, breeding both a connected society and isolated individual echo chambers where divergent views seldom meet to reconcile differences. The democratic potential of social media – to offer a voice to the voiceless – is threatened when AI-driven content algorithms overshadow diverse perspectives.

While the challenges are substantial, they are not insurmountable. Collaborative efforts between technologists, policymakers, and the public can yield frameworks that promote responsible AI usage in social media. Transparency in how algorithms function, alongside an emphasis on developing AI systems that celebrate diversity and fairness, will be vital. Efforts to educate the public about AI, misinformation, and digital literacy can empower users, allowing for a more informed digital citizenry.

Ultimately, the social implications of AI in social media call us to reflect on the nature of our interactions and the digital ecosystems we inhabit. These systems, woven intricately into the fabric of our daily lives, should enhance our capacity to connect and share authentically. How we navigate this evolving relationship with technology might very well define future societal structures and values.

Public Perception

In the burgeoning landscape of artificial intelligence, public perception hovers between awe and apprehension. For some, AI signifies a beacon of hope, promising a future of unfathomable innovations and convenience. Others, however, cast a wary eye, harbouring concerns over job displacement or the societal imbalance that might arise from unchecked technological advancement. Surveys and polls paint a complex picture, echoing an intricate tapestry of optimism and fear that underscores our collective consciousness. While enthusiasts marvel at AI's potential to revolutionise industries and solve global challenges, skeptics question its ethical ramifications and the readiness of regulatory frameworks to ensure its safe deployment. This dichotomy is a reminder that as AI continues to weave itself into the fabric of our daily lives, an informed public dialogue is essential to guide its integration into society responsibly.

Surveys and Polls serve as a compelling lens through which we can glimpse public sentiments about AI. Ever since AI started to seep into everyday life, from recommending your next binge-worthy series to helping diagnose medical conditions, how the public perceives it has become crucial. In exploring AI's social implications, asking the right questions can reveal a plethora of insights into societal attitudes, fears, and hopes.

Surveys provide a snapshot of what people think AI is capable of. For instance, a survey might reveal that a majority of people believe AI will vastly improve healthcare by catching diseases early. Yet, the same group might be sceptical about AI taking over creative tasks that are essentially human, like writing a novel or composing music. These polls highlight a fascinating tension between what people hope AI will achieve and the wariness of its ethical implications.

Polls also uncover regional differences in AI perception. In countries with rapid technological integration, uncertainties often give way to curiosity and optimism. In contrast, regions where job markets are volatile might display more apprehension, viewing AI as a harbinger of job displacement rather than opportunity. These geopolitical variations remind us that AI's reception isn't monolithic and is shaped by local contexts and experiences.

One of the intriguing outcomes from public opinion surveys is the ambivalence surrounding autonomous vehicles. An overwhelming percentage may express excitement about the potential decrease in accidents due to AI's quick decision-making abilities. Yet, the notion of relinquishing control remains a sticking point. Humans, it seems, aren't entirely ready to trust machines with matters of life and death, a reticence that's echoed in many other areas of AI integration as well.

Then there's the question of privacy. Polls frequently underscore public anxiety about how personal data is handled. While AI in social media offers personalized experiences, users are increasingly cognizant of how these platforms utilise their data. Many express unease over what AI knows about them, and surveys reveal a call for better transparency and stricter regulations.

Interestingly, surveys often reveal a generational divide in attitudes towards AI. Younger generations, who have grown up in a digital ecosystem, are generally more enthusiastic and trusting of AI technologies. They tend to welcome AI's presence as a seamless extension of their digital reality. Older generations might regard AI with more scepticism, shaped by a lifetime of understanding intelligence as inherently human.

The methodology of these surveys itself is worth noting. The way questions are phrased can influence outcomes, highlighting the subjective veneer underlying data collection. Asking, "How might AI improve your daily life?" could yield a torrent of optimistic responses. Conversely, framing the question as, "In what ways does AI threaten your privacy?" could unearth a trove of concerns. This duality speaks to the power of language in shaping perception and the importance of designing surveys that capture a balanced view.

Moreover, cultural narratives about AI continue to shape responses. Pop culture has long fed us visions of AI—often dystopian—that etch themselves into the social imagination. Whether it's the sentient AI uprising in films or the benevolent AI companion in literature, these stories influence how people envisage AI's role in society. Surveys tap into these underlying perceptions, revealing the stories people tell themselves about the future.

Another sentiment frequently captured is the hope that AI might solve some of the world's most persistent problems. Environmental issues, for instance, often surface in polls about AI's potential; people genuinely believe AI could herald breakthroughs in sustainable energy and climate monitoring. This optimism isn't just techno-utopianism but a reflection of the urgent need for innovative solutions to pressing global challenges.

Of course, for policy-makers and technologists alike, understanding public perception isn't just academic. It's instrumental in bridging the gap between AI ambitions and societal acceptance. Surveys provide a feedback loop, helping developers tweak technology to better suit the ethical and practical concerns of the people who use it. This feedback is invaluable, acting as a catalyst for more thoughtful, inclusive AI development.

Surveys and polls, then, are more than just numbers. They reflect a society in flux, negotiating the terms of a partnership with machines that have an extraordinary capacity for both change and innovation. They're signposts pointing the way to the societal landscapes AI might traverse as it continues to evolve. Above all, these surveys remind us that while technology advances relentlessly, it's the human element—our hopes, fears, and dreams—that will ultimately chart the course of AI's journey in society.

CHAPTER 15: THE FUTURE OF AI

The future of AI is a canvas of boundless possibilities, painted with the vibrant hues of emerging trends and groundbreaking research. As machine learning algorithms mature and computational power burgeons, we're on the cusp of an era where AI can truly augment human potential. Innovators are exploring frontiers that promise more intuitive interactions, seamless integration with daily life, and advancements that tackle global challenges—from climate change to healthcare. Speculation abounds about AI's role in society, raising questions about ethics and responsibility, yet fostering an optimistic anticipation of a future where AI collaborates rather than competes with us. The journey ahead isn't just about technological leaps but about harmonising our human values with machine logic. We stand on the threshold of a transformative age, ready to redefine the landscape of human endeavour with each step forward in AI development.

Emerging Trends

The world of artificial intelligence is in constant flux, with emerging trends painting a vivid picture of how the future might unfold. One of the most captivating trends is the rise of generative AI, where machines not only learn from existing data but also create novel content, crossing the threshold into realms like music, art, and literature. Meanwhile, AI is becoming more personalised, tailoring recommendations and experiences in ways that were once the domain of science fiction. The integration of AI with the Internet of Things (IoT) is another trend gathering momentum, as smart devices become even smarter, bringing the promise of unprecedented connectivity and automation into everyday lives. Furthermore, ethical AI is increasingly taking centre stage with advancements focusing on fairness and reducing biases, promising a future where technology harmonises more seamlessly with societal values. As we dive deeper into these currents, the exciting element isn't just the technological leaps themselves, but the myriad ways they might intertwine with human aspirations, pushing boundaries beyond our current imagination.

AI Research Frontiers In this dynamic landscape of technology, the sphere of artificial intelligence is not just about what it is now, but also about where it's heading. Imagine a world where machines understand not only language, but context and emotion. They don't just interact with humans but also anticipate their needs and responses in ways once considered science fiction. This notion, although ambitious, is at the heart of current AI frontiers.

The potential for AI to revolutionise our daily lives has intrigued scientists and technologists alike, pushing them to explore boundaries that were previously uncharted. Among the most exciting areas is the development of human-computer interaction that feels as natural as conversations between people. This involves advanced natural language processing models that transcend mere semantic comprehension and begin to understand the subtleties and nuances of human emotion and intention.

One significant area of exploration is the enhancement and scaling of machine learning models. Researchers are constantly working towards creating algorithms that are not only more efficient but can also learn from smaller datasets. This would mean that even in data-scarce environments, AI systems could be effectively trained, opening up possibilities for applications across varied and previously inaccessible domains. Existing innovations in transfer learning and federated learning exemplify how AI is moving towards models that can be more universally applied, learning from minimal data while maintaining privacy.

Neural networks, particularly deep learning architectures, have captured our imaginations due to their ability to perform complex tasks with remarkable success. However, the journey has just begun. Advances in neural network design are enabling models to become more interpretable and less of a "black box," addressing one of the major criticisms levelled against AI systems. Researchers are also exploring neuromorphic computing, which seeks to mimic the human brain's architecture and processes, promising substantial increases in processing power and efficiency.

Another frontier lies in the development of explainable AI (XAI). As AI systems become more embedded in decision-making processes, the necessity for models that can be interpreted and understood by humans becomes paramount. Explainable AI isn't just a luxury—it's a necessity for ensuring trust and accountability in AI-driven decisions. The drive for transparency is steering research into methods where algorithms can elucidate their decision pathways, thus fostering a greater degree of reliance from users.

The ethical dimension of AI research is equally vital. With AI's expanding capabilities, ensuring ethical considerations are front and centre is crucial. Research is being undertaken to create models that are unbiased and fair, addressing criticisms about algorithmic discrimination. By integrating ethical frameworks into the design process, researchers aim to mitigate the risk of perpetuating societal biases within AI systems.

In the realm of AI's potential impact on global challenges, research is pivoting towards creating solutions for pressing issues like climate change, health crises, and inequality. AI systems capable of predicting environmental changes, optimising renewable energy usage, and even contributing to global health by predicting disease outbreaks or tailoring personalised medicine are at the forefront of this research frontier. Such advances promise not only new technological tools but a better quality of life and a more sustainable planet.

As AI systems evolve, so does the need for robust infrastructure to support these burgeoning technologies. Research into quantum computing represents this frontier, with its potential to exponentially boost computational power. By harnessing the principles of quantum mechanics, researchers aim to overcome the limitations of classical computing, opening the door to solving complex problems that were previously deemed unsolvable.

The field of robotics, too, is undergoing significant transformations. The integration of AI with robotics is leading to machines that can perform tasks with an unprecedented level of autonomy and dexterity. From caregiving robots in healthcare to autonomous drones in agriculture and logistics, AI-driven robotics stands to fundamentally alter the way industries operate and interact with the natural world. Researchers are focusing on making these systems more adaptable and resilient, allowing them to operate effectively in diverse and changing environments.

Moreover, the boundaries of AI research are extending beyond Earth. Space exploration, with its unique challenges, is increasingly relying on AI for mission planning, data analysis, and even autonomous decision-making. Robotic explorers are being equipped with AI-driven capabilities that allow for greater autonomy, supporting humans in these far-reaching missions. Researchers are laying the groundwork for AI that can one day aid in establishing human colonies on other planets.

We stand on the cusp of a new era, where the frontiers of AI research encourage a vision of technology that collaborates with humanity rather than conquers it. Harnessing AI's potential in ways that are harmonised with ethical considerations and human values is not only desirable but essential. As we venture further into the depths of AI exploration, staying true to our goal of responsibly navigating and shaping this technology will create pathways where AI benefits society at large.

Predictions and Speculations

The landscape of artificial intelligence is ever-evolving, and while we marvel at the achievements we've uncovered so far, one can't help but cast an eye toward the shimmering horizon of what's to come. Understanding AI's future involves placing informed bets on emerging trends, exploring speculative, yet scientifically grounded possibilities, and envisioning societal transformations so profound they might make sci-fi narratives seem quaint.

One key trend anticipated in the AI realm is the accelerated pace at which machine learning models learn and adapt. Future iterations of AI are expected to develop a level of autonomy in learning that's beyond our current methods. Imagine AI systems that don't simply learn from the vast datasets we provide but actively seek out new data, testing hypotheses and refining their own algorithms independently. This self-improvement cycle could lead to rapid advancements in AI capabilities, dramatically shortening development cycles and ushering in innovations at a never-before-seen pace.

Moreover, as AI technologies continue to permeate various facets of our lives, a more profound integration of AI systems into the fabric of society seems inevitable. Emerging applications won't just enhance individual industries; rather, they might redefine what it means to be a part of these sectors entirely. The concept of AI as a silent partner, seamlessly working within our everyday operations, is poised to become the norm. From tailored healthcare solutions birthed by algorithms predicting individual wellness needs, to AI-driven finance models that can anticipate and counteract economic fluctuations, the scope is as extensive as it is thrilling.

We also find ourselves on the brink of a significant shift in how we interact with AI. Today, virtual assistants are helpful yet limited; however, future AI systems are likely to develop conversational interfaces so sophisticated that interacting with them could become indistinguishable from engaging with another human. This evolution in human-AI interaction could rejuvenate sectors like education, where AI tutors offer more personalised, engaging, and effective learning experiences. In the workplace, conversational AI might transform collaboration, bridging cultural and linguistic divides to foster a truly global workforce.

A discussion about the future of AI invariably leads to the tantalising prospect of artificial general intelligence (AGI). The idea of machines possessing human-like understanding, reasoning, and perception has long fired the imaginations of scientists and philosophers alike. Although today's AI predominantly excels in narrow, specialised tasks, many experts predict that achieving AGI is not a matter of "if," but "when." The road to AGI is fraught with technical challenges and ethical quandaries, yet each stride taken towards it brings with it a utopian vision where AI functions as an equivalent partner in human endeavours, propelling humanity into a new era of enlightenment.

Of course, no exploration of AI's future would be complete without contemplating its ethical and societal impacts. Predictions suggest an augmented focus on developing ethical guidelines that govern AI deployment, ensuring that the technology benefits all of humanity, not just a privileged few. There's a strategic movement towards embedding ethical decision-making frameworks within AI systems themselves, allowing these entities to make choices aligned with societal values and human rights.

The potential of AI to exacerbate or ameliorate existing inequalities will also come to the fore. As AI technology spreads, it holds both the promise of narrowing existing global divides and the risk of widening them. Equitable access to AI resources and education will be essential in creating balanced growth across diverse populations. Efforts to bridge the digital divide will likely intensify, striving to create a future where technology serves as a leveller, not a divider.

Speculations around AI's influence on employment present both positive and negative forecasts. While AI has the potential to automate mundane jobs, it's also poised to create a plethora of new opportunities that we're only beginning to imagine. The birth of entirely new industries, job roles centred around the ethics, management, and maintenance of AI, and a shift in job focus from hard skills to creativity and interpersonal skills are just a few of the possibilities on the employment horizon.

Furthermore, AI's impending transformations extend beyond tangible improvements, potentially sparking a profound cultural shift. The concept of intelligence and creativity might evolve, with traditional human activities, like music composition or painting, experiencing reinvention through AI collaboration. Societal perceptions of what is art or creation may expand, challenging preconceived notions of creativity entirely.

As we edge further into the potential decades ahead, it's crucial to remain vigilant about AI's environmental impact. The data-driven approach inherent to AI consumes significant energy resources, and predictions indicate a growing focus on sustainable, eco-friendly AI development. Carbon-neutral data centres, energy-efficient algorithms, and AI-driven solutions to environmental conservation are likely to form the bedrock of future AI endeavours.

With each new breakthrough, AI promises to reshape the world in ways that defy our current understanding. These predictions and speculations serve not just as a roadmap for technological progression but as a mirror reflecting our aspirations and anxieties about technology's role in our future. While the specifics of this future remain uncertain, the promise of AI, as a partner in our quest for progress and innovation, remains undeniably exciting.

CHAPTER 16: AI AND CREATIVITY

In the sphere of creativity, AI has begun to carve out a niche that's as fascinating as it is transformative. Imagine a world where algorithms paint pictures, compose symphonies, and draft compelling narratives, all while collaborating with humans to push the boundaries of art and design. AI isn't just a tool; it's becoming an active player in artistic endeavours, fueling innovation by breaking down the barriers of traditional creative processes. Whether it's AI-generated music that's reshaping the music industry or machines contributing to intricate design work, the implications for creativity are profound. Artists and writers are increasingly drawn to this new collaborator, finding in it not a competitor, but a catalyst for inspiration, a way to explore the unexplored and reinvent the familiar. As AI continues to evolve, the symbiotic relationship between human and machine promises to redefine what creativity means in the digital age, envisioning new forms of expression and storytelling that were once confined to the realm of science fiction.

Art and Design

In the vibrant realm of art and design, artificial intelligence is shaking the ground beneath the feet of traditional creativity, ushering in a new era where machines are not just tools but creative partners. Imagine algorithms that can generate paintings, designs, or even fashion pieces with a touch that's almost, but not quite, human. This world isn't far removed from the one we know; rather, it's an expansion, adding layers of complexity and nuance to artistic endeavours. Artists now exploit AI's ability to sift through vast datasets, uncovering patterns and inspirations that might elude human perception. Designers collaborate with AI to innovate in ways that are efficient yet brimming with originality. While some sceptics view these advancements warily, fearing the loss of the artist's unique touch, the infusion of AI into art and design is often more a symbiosis than a replacement. The creative process becomes a dance, one where machine learning and human intuition intermingle, redefining what we envision possible, and ultimately pushing the boundaries of imagination itself.

AI as an Artist marks a new chapter, not just in technology, but in the very essence of what it means to create. Imagine a process traditionally bound by imagination, the hum of a rotating disc, the soft stroke of a brush meeting canvas, and even a crescendo in a symphony, now embedding AI as a possible collaborator. In recent years, AI has begun not only to perform tasks that imitate artistic styles but to push the boundaries of artistic expression itself. Is this collaboration or competition? It's a conundrum that echoes across both artistic and tech communities, sparking curiosity and debate. AI as an artist challenges the notion of creativity, touching realms we previously thought uniquely human.

In the history of art, creativity was often considered a divine gift, a spark that separated humans from machines. However, machines have started to learn and detect patterns, making their foray into the artistic world. AI, through complex algorithms and neural networks, analyses existing artworks to understand essence, style, and innovation. By doing so, it doesn't just learn to replicate but learns to innovate. The question stands: Can a machine be truly creative, or is it simply regurgitating data in aesthetically pleasing forms? When contemplating these questions, consider how AI has successfully created artworks imitating the surrealism of Dali or the impressionist strokes of Monet, sometimes indistinguishably so.

We must look deeper, especially at notable projects like Google's DeepDream, which takes neural network layers designed for image recognition and repurposes them for creating hallucinogenic art. It demonstrates the potential of AI to transform what's mundane into something replete with abstraction and eeriness. This shows that AI doesn't merely copy but creates unique interpretations, reflecting a kind of artistic sensibility once thought to be inaccessible to machines.

There's a growing presence of AI-driven artwork in prestigious galleries and auctions around the world. In 2018, an AI-generated artwork, "Portrait of Edmond de Belamy," made headlines when it sold at a major auction house for an astounding $432,500. This remarkable event signalled the art world's acceptance of AI-created art, placing machine intelligence alongside renowned human artists. AI-produced art isn't just a manifestation of technique—it's about perspective, challenging us to reassess our definitions of creativity and art.

Let's explore collaboration too. Beyond creation, AI serves as a tool, augmenting and enhancing human creativity. AI applications assist artists in discovering new styles by combining and contrasting different works, suggesting palettes that might evoke specific moods or themes. This art-tech synergy doesn't steal the limelight from human creativity but rather provides artists with new avenues to express their vision. For example, in music, AI algorithms help composite efficiently, suggesting harmonies and offering rhythm alternatives, blending tradition with futuristic sounds.

There is no denying the scepticism surrounding AI's artistic involvement. Critics argue that true artistic creation comes from human emotions, experiences, and philosophies which a machine doesn't possess. However, this discourse prompts us to assess whether art must always convey emotion or can exist separately, as a pure form of expression. AI's role in art provokes a reimagining of creativity, leading us to reflect on the intent, sentiment, and authenticity of artworks in our digital age.

The blend between AI and artist doesn't end in creation alone. In the realm of design, AI offers significant contributions. Graphic design, for instance, sees AI engines proposing novel layouts or assisting in logo creation based on specific themes or business identities. Fashion design also benefits, with AI algorithms predicting trends, suggesting fabric patterns, or even crafting custom garments that strike the perfect balance between comfort and avant-garde styles.

There's this intrinsic curiosity about whether AI's ability to create art might ever equate to the human touch, leading to philosophical inquiries about consciousness and abstraction. Even as machines learn more about the nuances of artistic creativity, the nuances of what makes us human keep shifting. In the end, it could be that AI might never fully replace the human artist, but serve as an exponential tool for human creativity, challenging the boundaries, breaking the norms.

As we navigate this new terrain of AI artistry, the narrative isn't just about robots mimicking human intellect but about involving AI in the co-creation of art, music, and design—an inclusive experience where technology complements, rather than competes with, human intelligence. These technological advancements redefine our conception of art, opening doors to infinite possibilities in creativity. AI as an artist or partner signals change, a possibility that is as exciting as it is uncharted, forging a path towards creativity unbound by traditional constraints.

As technology enthusiasts, creatives, or casual admirers of art, we're invited to journey into uncharted territory, considering AI not just as a tool but as a brave new medium. The fusion of human ingenuity with machine precision might just lead us to uncover new depths of imagination, an exploration of art in dimensions yet imagined. The symphony of human and AI-crafted art has only begun its overture, promising a world where the lines of creativity continue to blur and expand.

Literature and Writing

When discussing "Literature and Writing" in the realm of AI and creativity, we venture into a captivating dialogue about the symbiosis between human ingenuity and machine precision. Imagine algorithms that don't just understand language but spin stories that intrigue, educate, and entertain. As AI begins to craft narratives, weave plots, and develop characters, it prompts us to question the very essence of creativity. Are these digital scribes simply extensions of human creativity, or do they herald a new epoch of storytelling? The marriage of AI with literature challenges traditional notions of authorship while offering novel possibilities—co-authored stories and adaptive narrative arcs conscious of reader engagement. This fusion doesn't diminish the role of the human writer but expands the horizons of what's possible, inviting creators to explore uncharted territories of collaborative storycrafting. In this interplay, AI becomes not just a tool, but an ally in the eternal pursuit of expression and connection through the written word.

AI in Storytelling is an evolving frontier that brings profound changes to literature and writing. As AI technologies become more sophisticated, they are not just tools for refining human-written narratives; they are also creators in their own right. Imagine a digital scribe that churns out sonnets or produces compelling narratives, intriguing both readers and critics. These AI-driven stories blend human-like creativity with the precision and logic of machine algorithms, offering a new kind of literary experience.

For decades, writers have explored the idea of AI-generated stories, often depicted in speculative fiction as cold and mechanical. However, contemporary AI in storytelling is anything but monotonous. It can mimic the styles of famous authors or develop unique voices that defy traditional norms. How does it accomplish this? Through machine learning algorithms that analyse vast amounts of text, enabling AI to understand narrative structures, character development, and even emotional nuances. As it learns, it becomes capable of crafting stories that entertain, provoke thought, and resonate emotionally with readers.

The implications of AI in storytelling extend beyond producing text. It impacts the publishing industry, offering new possibilities for writers who might otherwise struggle to find a platform for their work. With AI tools that assist in plot creation, character development, and thematic exploration, aspiring authors can explore narrative ideas more freely. This democratisation of creativity enables storytellers to experiment without the financial constraints traditionally associated with publishing.

Yet, there lies an intriguing paradox in AI storytelling. While it offers fresh insights and enhances human creativity, it also raises questions about authorship and originality. Who owns a story written by AI? Is it the programmer who designed the algorithm, or the AI itself? Perhaps the real authorship belongs to the collective knowledge encapsulated in the dataset used for training. These are not just philosophical queries; they have real-world implications for copyright laws and intellectual property rights. As our understanding of AI storytelling evolves, so too must our legal frameworks.

Not all reactions to AI storytellers are enthusiastic. Detractors argue that AI lacks true creativity and that its stories are mere pastiches of human-written works, devoid of genuine artistic merit. They assert that creativity is a distinctly human trait, grounded in the unique experiences and emotional complexity of individuals. While AI can certainly mimic pattern recognition and logical construction, can it really grasp the essence of human experience? In response, proponents point out that AI is designed not to replace human creativity but to augment it. AI can push the boundaries of imagination, sparking new directions that human writers might not envision independently.

As AI continues to infiltrate the world of storytelling, it also shapes the way stories are consumed. Personalised narrative experiences emerge as one of the most exciting developments. Imagine reading a novel that adjusts its plot, setting, or characterisation based on your preferences, emotional responses, or even your mood at the time. Machine learning algorithms can analyse reader engagement and adapt storytelling elements in real-time, creating a bespoke literary journey tailored to individual readers.

In educational contexts, AI enhances the writing and learning process. Students receive tailored feedback on their writing, with AI-driven platforms acting as personal tutors. These systems can provide immediate suggestions for improving grammar, style, and narrative coherence. AI's capability to offer constructive criticism and even suggest alternative narrative paths encourages students to think critically about their work, fostering a deeper understanding of storytelling mechanics.

The future of AI in storytelling holds immense potential but also requires a thoughtful approach. Developers and writers should strive for synergistic collaborations that respect the art of storytelling. This involves designing AI models that appreciate cultural sensitivities, ethical considerations, and the intrinsic values that literature and writing uphold. As creators embrace AI, the focus should remain on harnessing technology to enrich human narratives rather than dilute them.

Ultimately, AI in storytelling reflects a larger conversation about creativity in the digital age. It challenges preconceived notions about what it means to be a writer and what constitutes a story worth telling. Will AI crafting narratives lead to a renaissance in literature, or might it drown human voices in a sea of algorithmically generated text? The answer lies in how we choose to leverage these technologies, balancing the magic of human imagination with the computational power of AI.

CHAPTER 17: AI AND CLIMATE CHANGE

As we venture into the intricate dance between AI and climate change, it's clear that technology is stepping up as a guardian of our planet's future. AI is transforming how we monitor and predict environmental shifts, offering unprecedented precision in tackling unpredictable climate patterns. Predictive models powered by machine learning are not just forecasting weather; they're becoming critical tools in climate science, enabling quicker responses to natural disasters. Furthermore, AI is optimising renewable energy systems, making them smarter and more efficient. This technological symbiosis enables better allocation of resources, reducing waste, and promoting sustainable practices. Imagine solar panels harnessing maximum sunlight or wind turbines adapting in real-time to changing weather conditions. As AI breathes intelligence into our environmental efforts, it inspires a future where innovation meets conservation, crafting a narrative of balance and sustainability.

Environmental Monitoring

Environmental monitoring powered by AI has become an invaluable asset in our quest to tackle climate change. These advanced systems harness vast amounts of data from satellites, sensors, and other sources to track and analyse environmental changes with astonishing precision. AI models can detect subtle shifts in temperature, vegetation cover, and pollution levels, providing timely insights into the health of our planet. Such capabilities allow for more accurate predictions of extreme weather events, enabling better preparation and response strategies. The integration of machine learning algorithms with environmental data not only enhances our understanding of ecological dynamics but also inspires innovative solutions for conservation efforts. As we continue to face the challenges posed by a changing climate, AI stands as a beacon of hope, offering novel ways to protect and sustain our natural world.

Predictive Models have become the unsung heroes of environmental monitoring in our fight against climate change. As the planet grapples with the multifaceted challenges of changing climates, predictive models offer a glimpse into the future, empowering us to make informed decisions today. At the core, these models are not just complex algorithms but rather, they're intricate maps guiding us through the labyrinth of environmental possibilities and perils.

Predictive modelling is akin to peering into a crystal ball—though far more reliable. They leverage vast data streams, ranging from satellite images to sensor readings, to forecast weather patterns, evaluate climate variations, and predict natural disasters. Imagine a world where we can anticipate the path of a hurricane with pinpoint accuracy or prepare for a drought months before it strikes. With AI-enhanced predictive models, these scenarios shift from the realms of imagination into reality.

In essence, predictive models are all about synthesis. They take disparate shards of information—temperature readings, ocean currents, air pressure—and weave them into cohesive narratives that tell us what may lie ahead. AI's ability to process and analyse colossal datasets in real-time means we're no longer in the dark about immediate environmental changes. Instead, we possess the tools to forecast and adapt, potentially saving countless lives and ecosystems in the process.

One of the significant breakthroughs in predictive modelling is the ability to identify patterns invisible to the human eye. Consider ocean currents, for example. Traditional methods might miss subtle shifts in these massive water bodies. Yet, predictive models equipped with machine learning can discern these minute alterations, offering insights into potential climate impacts. Such capabilities are redefining how we understand the very forces shaping our world.

Moreover, predictive models are proving to be invaluable in mitigating the effects of climate change on agriculture. By predicting weather conditions, farmers can adjust their planting and harvesting schedules to better align with natural phenomena. This foresight is crucial not only for safeguarding crops but also for ensuring food security amid a changing climate. Essentially, we're crafting an agricultural ecosystem that thrives by anticipating and countering the whims of nature.

Beyond agriculture, predictive models play a crucial role in water resource management. As global populations surge and water become scarcer, the need to efficiently manage this precious resource becomes paramount. Predictive analytics can forecast periods of water scarcity and abundance, allowing for better reservoir management and water distribution strategies. These models arm policymakers with the data needed to implement preemptive measures, thereby averting potential humanitarian crises.

In the realm of disaster preparedness, predictive models have transformed our capabilities profoundly. By simulating possible scenarios of natural disasters such as floods, tornadoes, or wildfires, these models help in planning and executing swift response strategies. They provide critical data that emergency services can use to position resources effectively, thereby reducing the impact on human lives and property. This proactive approach marks a shift from reactionary disaster management to a more anticipatory stance.

However, it's not just about the big-picture projections; predictive models offer granular insights as well. They can inform local governments about air quality variations within cities, helping to implement smarter city planning and pollution control measures. Given the challenges of urbanisation and industrialisation, these models offer a way to balance economic growth with environmental stewardship, ensuring that urban areas develop sustainably without compromising air quality.

As inspiring as these possibilities are, they come with their complexities. The effectiveness of predictive models hinges on the quality and granularity of data fed into them. Any inaccuracies or biases in data can ripple through the model, producing misleading forecasts. Thus, constant refinement and validation of these models are crucial for maintaining their reliability and effectiveness. This requires an ongoing dialogue between data scientists, environmental experts, and policy-makers to ensure that these models remain robust and tailored to current realities.

Additionally, there's an ethical dimension to consider. Developing predictive models necessitates access to vast amounts of data, often raising privacy concerns. Striking the right balance between accurate predictions and individual privacy is a nuanced challenge that requires careful navigation. Transparency in how data is collected and used is vital to fostering public trust and ensuring compliance with ethical standards.

Another evolving aspect is the accessibility of predictive technology. To truly combat climate change, these tools need to be available to all nations, not just the most resource-rich ones. This democratization of technology ensures that even the most vulnerable communities, often disproportionately affected by climate change, have the wherewithal to forecast and prepare for potential environmental challenges.

In conclusion, predictive models stand at the frontier of AI and climate change, offering a beacon of hope amid uncertainty. They embody a shift towards data-driven decision-making, informing every step we take in safeguarding the planet's future. By integrating AI's predictive prowess with human ingenuity, we set the stage for a sustainable future—one where our responses to environmental challenges are not merely reactive but thoroughly premeditated and finely tuned to the nuances of our ever-evolving world.

Renewable Energy

In the realm of renewable energy, AI acts as a transformative force, driving efficiency and innovation. As solar, wind, and other green sources continue to fight climate change, AI offers potent tools to maximise their potential. With the ability to analyse vast datasets quickly, AI systems can predict energy production and consumption patterns, optimising the grid for peak efficiency and reducing waste. It balances load demands by forecasting weather conditions more accurately than ever before. Imagine solar panels and wind turbines that communicate with AI-driven systems, adjusting outputs dynamically to match real-time demand in the electrical grid. Such synergy between AI and renewable energy technology not only enhances sustainability but also accelerates the transition towards a carbon-neutral future. It's not just about generating clean energy; it's about intelligently distributing it, ensuring that every joule is used to its maximum potential. By streamlining operations and improving the reliability of renewable sources, AI injects new life into global efforts to combat climate change, proving that technology doesn't merely serve humanity's needs but can also safeguard its environment.

Optimising Efficiency You've no doubt heard about how AI is revolutionising nearly every industry, but when it comes to renewable energy, the potential is downright transformative. Technologies like wind, solar, and hydroelectric power are already beneficial for the planet, yet integrating AI can supercharge their effectiveness, helping us make smarter use of natural resources. Furthermore, AI can fine-tune these energy systems, allowing them to operate at peak efficiency by using algorithms that predict, monitor, and adjust in real-time.

Imagine a wind farm operating at suboptimal levels because of a shift in wind patterns or minor equipment inefficiencies; here enters AI to save the day. AI algorithms can predict fluctuations in wind patterns and adjust turbine operations accordingly, maximising output. The blend of AI and sensors in wind farms creates smart systems that continually learn and adapt, boosting the efficiency of power generation. AI is also capable of predicting maintenance needs, allowing for timely interventions and reducing downtime—beneficial for both the pocket and the planet.

In solar power, the optimisation possibilities are equally astounding. AI models can analyse historical data and real-time weather information to predict energy output and maximise the positioning of solar panels. You may wonder why positioning is crucial. Well, incorrect angles can lead to underperformance, something AI can entirely circumvent, ensuring that solar arrays capture every possible photon. As if that's not enough, AI can also help in predictive maintenance of solar panels, keeping them free of dust and debris, and scheduling regular check-ups to prevent efficiency drops.

Many energy providers are now leveraging AI to tackle the significant problem of energy storage. Energy storage systems like batteries are essential for managing supply and demand. AI can facilitate more effective storage solutions by predicting demand patterns and adjusting accordingly. Think of AI as the brain behind the storage, figuring out when to store energy and when to release it, all the while keeping costs as low as possible. This not only enhances efficiency but also ensures a stable energy supply on cloudy or windless days.

Grid management is another area ripe for AI intervention. Traditional grids often struggle with energy distribution inefficiencies, especially when introducing renewable energy resources. AI improves grid operations by predicting energy demand, enabling operators to better balance loads and reduce waste. Through machine learning models, AI systems create predictive technologies capable of forecasting energy generation and consumption patterns. This allows for more efficient distribution of energy across the grid, minimising losses and optimising resource allocation. In addition, AI can swiftly adapt to faults or sudden changes, maintaining efficiency even under unpredictable conditions.

You see, AI is not just about creating efficiencies in the production and storage of renewable energy; it's also about refining energy consumption at the customer end. AI can assist smart homes in learning power consumption patterns and automatically adjust settings like thermostat levels, lighting, and appliance use to reduce waste. When multiplied across thousands of homes and businesses, such optimisations can lead to significant energy savings.

Now think logistics and transportation—another realm where renewable energy can leap ahead with AI. Electric vehicles, for instance, rely on optimised algorithms to extend battery life and improve charging efficiency. Batteries themselves benefit from AI by predicting the best charging routines, reducing wear and tear, and consequently lowering lifetime costs. As more vehicles hit the road, AI's role in managing charging infrastructures and ensuring that grid load is evenly balanced will be pivotal.

Moreover, local governments and corporations are recognising how AI and automation can drive change in urban planning, adopting sustainable energy practices that favour low-carbon footprints. Digital twins—virtual models of physical assets, systems, or entire cities—can simulate real-world scenarios, enabling architects and engineers to experiment with renewable technologies and optimise infrastructure before a single brick is laid.

Nations are also leveraging AI to refine and perfect policy frameworks guiding renewable energy adoption. Policymakers can harness AI-driven data analytics to evaluate regulatory impact in real-time, paving the way for more adaptive and effective frameworks that encourage clean energy usage. Efficient governance married with AI-generated insights can tangibly hasten the adoption of renewable resources.

Still, let's not ignore the challenges. Integrating AI into renewable energy sectors comes with its own set of hurdles, primarily relating to data management, interoperability, and upfront costs. But as AI scales, these challenges are being tackled with innovation and investment. New startups are entering the space, dedicated to specific niches, like deploying AI for microgrid management or perfecting machine learning models that outsmart today's inefficiencies. As these solutions gain traction, economies of scale come into play, further enhancing efficiency opportunities at lower costs.

While we're here discussing optimisation, it's crucial not to fall into the trap of relying on AI as a panacea. AI is a tool—a powerful one—but one that must operate within a well-structured framework. Data accuracy and ethical considerations are vital. An extraordinary amount of data drives these AI systems, and any error can lead to mismanagement rather than optimised results. Thus, the role of humans in overseeing AI technologies ensures that we're always steering them in the right, ethically sound direction.

In summary, integrating AI into renewable energy systems opens up a realm of optimisation that was unimaginable a few decades ago. Every node in the chain, from production and storage to consumption and management, stands to benefit from what AI offers. As AI continues to mature, expect even greater efficiencies and deeper integrations into our power systems, reinforcing the critical need for human oversight to ensure that this technology serves societal needs as sustainably as possible. The future is not merely technology-driven but smartly orchestrated, aligning with the ongoing quest for a truly sustainable planet.

CHAPTER 18: AI AND SECURITY

As artificial intelligence continues to evolve, its role in security can't be understated, traversing the thin line between innovation and vulnerability. Cybersecurity sees AI as both a shield and a sword, adept at identifying incursions faster than the human eye could spot, yet also being a potential tool for orchestrating more sophisticated cyber attacks. On the national front, the integration of AI into defense strategies promises enhanced surveillance and predictive capabilities, yet this advancement requires a careful balance between technological ambition and ethical constraints. The dream of an AI-driven secure world mirrors the complexity of a digital landscape where the boundaries of privacy and security are constantly being tested. Harnessing AI's power in security beckons with promises yet demands vigilance, lest the very systems designed to protect us fall prey to the cunning they are meant to curtail.

Cybersecurity

In the constantly evolving landscape of AI, cybersecurity emerges as a keystone in protecting both data and infrastructures from increasingly sophisticated threats. Picture AI as both a guardian and an adversary; it fortifies our digital borders with its unparalleled ability to detect patterns and anomalies, making it instrumental in threat detection and prevention. Yet, with this power comes a new wave of cyber threats, as malicious actors also harness AI to launch more effective and elusive attacks. This cat-and-mouse game paves the way for a dynamic evolution of cybersecurity measures, where AI-driven tools are not just a luxury, but a necessity. As technology enthusiasts and professionals, the challenge lies in staying one step ahead, crafting systems that learn and adapt faster than the threats they are designed to thwart. It's a realm where human ingenuity intersects with machine efficiency, forming a crucial alliance to safeguard our digital existence.

Threat Detection in the realm of cybersecurity intertwines itself with AI in a symbiotic dance, crafting a shield of perpetual vigilance in the digital landscape. In an era where cyber threats morph at an unprecedented pace, AI-driven threat detection emerges as a protagonist, offering both speed and precision that simply outpace traditional methods. The melding of AI and cybersecurity offers a foresight, predicting potential vulnerabilities and recognising threats in real-time.

AI in threat detection isn't just about automation; it's about transformation. By leveraging machine learning algorithms, AI systems analyse vast datasets to unearth patterns that hint at malicious activity. These patterns could include unusual login behaviours or unexpected data transfers that humans might overlook. The result is a proactive approach to cybersecurity where threats can be neutralised before they inflict damage. It's a bit like having a digital immune system that's constantly updating itself, learning from each new encounter.

This shift towards AI-enhanced cybersecurity offers a dual advantage: scalability and adaptability. Human analysts, no matter how skilled, have limits on how much data they can process—we're talking about millions of potential threats daily. AI, however, thrives in this environment, scaling effortlessly to analyse a breadth of data that was previously unmanageable. Its adaptability comes from machine learning; as new types of cyber attacks surface, AI systems evolve, refining their algorithms to recognise and mitigate these threats effectively.

The journey of AI in cybersecurity isn't without its challenges though. One significant hurdle is the potential for false positives, where legitimate activities are flagged as threats. This can lead to a sort of alert fatigue, where security analysts become overwhelmed with constant notifications of potential threats that turn out to be benign. Bridging this gap requires a delicate balance—a combination of AI precision and human intuition. Incorporating AI into cybersecurity isn't about replacing the human element but augmenting it, creating a partnership that is greater than the sum of its parts.

The potential for AI in threat detection is practically boundless. Imagine a world where security systems predict and combat not just existing threats but anticipate future ones, analysing data trends to foresee cyber threats that haven't even been conceived yet. This horizon isn't a distant dream but an evolving reality. Technology enthusiasts and professionals are witnessing AI systems that don't just react; they simulate, predict, and preempt. This foresight is particularly crucial in sectors such as finance and healthcare, where the cost of security breaches can be catastrophically high.

Moreover, AI's capacity to work around the clock cannot be overstated. Cyber threats don't adhere to conventional working hours, and neither do AI systems. By providing relentless monitoring 24/7, AI ensures that there's never a moment when defences are down. This continuous vigilance is a formidable deterrent to cyber criminals, who may look elsewhere for easier targets.

AI-driven threat detection also ushers in the era of user-centric security. Adaptive security mechanisms personalise defence protocols based on user behaviour and habits. This adaptability means AI can distinguish between normative and anomalous behaviours, allowing for tailored responses to threats. The user experiences a seamless interaction, with robust security measures that don't intrude or disrupt.

AI, in many ways, is simplifying the complex world of cybersecurity by making it more dynamic, responsive, and user-friendly.

The ethical considerations of AI in threat detection aren't trivial either. As AI systems absorb and analyse data, privacy concerns rise to the surface. The challenge lies in striking a balance between deep analysis needed for threat detection and maintaining individual privacy. Data must be handled with utmost care, ensuring compliance with privacy regulations while still reaping the benefits of AI in cybersecurity. These ethical challenges, while daunting, are pivotal to fostering trust in AI-driven systems.

Furthermore, the exponential growth of IoT devices creates another frontier where AI can shine. These devices, interconnected and ubiquitous, offer convenience at the potential cost of security. AI's ability to manage and secure a vast ecosystem of devices – each a potential entry point for cyber threats – cannot be underappreciated. In essence, AI acts as both the watchtower and nurturer of this internet-based ecosystem, ensuring that each device harmoniously yet securely coexists.

It's not just the corporate or governmental sectors benefitting from AI in threat detection. Individual users, too, are reaping the benefits. Personal data breaches have, in recent years, highlighted the need for personal cybersecurity. Utilising AI-based solutions on personal devices helps detect phishing attempts or malware before they breach the defences, providing peace of mind to users navigating the digital realm.

In conclusion, AI's role in cybersecurity, particularly in threat detection, represents an exciting convergence of technology and foresight. As AI continues to evolve, so too will its capacity to protect, providing digital sanctuaries in an ever-changing cyber landscape. This changing dynamic urges technology enthusiasts, professionals, and society at large to embrace the potential of AI in crafting a safer, more secure digital future. In doing so, AI doesn't just detect threats—it transforms the very architecture of digital security, paving the way for a resilient technological era.

National Security

In the realm of national security, artificial intelligence emerges as both a formidable ally and a potential adversary. As countries worldwide harness AI to boost defensive capabilities, the stakes for military and intelligence operations escalate dramatically. AI's ability to process vast amounts of data in real-time can enhance threat detection, allowing for swift responses to potential risks. Imagine the fusion of machine learning with surveillance systems offering unprecedented situational awareness, or predictive analytics foreseeing geopolitical shifts before they unfurl. Yet, with these advancements come new vulnerabilities: autonomous weapons raising ethical quandaries and the race for AI dominance inciting a modern arms competition. The delicate balance of employing AI in national security underscores the need for robust, ethical frameworks to ensure these powerful tools bolster, rather than compromise, global stability and peace.

AI in Defence As the world shifts increasingly towards advanced technologies, AI emerges as a cornerstone in the defence sector, fundamentally altering the landscape of national security. This transformation isn't just a glimpse into a science fiction novel—it's happening now and it's redefining warfare as we know it. Military forces worldwide are adopting AI to enhance their capabilities, improve decision-making processes, and gain an edge over potential adversaries. With AI systems advancing rapidly, there's an pressing need to explore how these technologies integrate into defence operations and what ethical challenges they might present.

At the heart of AI in defence is the development of autonomous systems, designed to operate and make decisions independently of human intervention. These systems include drones, vehicles, and even robots capable of performing tasks that were once solely within the human domain. These autonomous entities can cover a wide range of operations, from surveillance and reconnaissance missions to precision strikes on target locations. Tanks that move autonomously across battlefields or naval vessels that patrol waters without human crews onboard seem futuristic. Yet, they're becoming the pulse of modern militaries, reshaping tactics and strategies on the ground, at sea, and in the air.

Machine learning and data analytics also play vital roles in defence by providing critical insights from vast amounts of data. For instance, AI-powered systems can analyse intelligence reports, satellite imagery, and other data sources to predict potential threats and recommend strategic actions. These tools enable military personnel to sift through information more efficiently and make informed decisions faster than traditional methods allow. Imagine the advantage of knowing your adversary's next move before they make it, thanks to predictive analytics that churns through terabytes of data in mere seconds.

In the domain of cybersecurity, AI acts as both a shield and a sword. Defensive AI systems can detect and respond to cyber threats in real-time, safeguarding critical infrastructure against digital attacks. By learning from patterns and anomalies, these systems can identify potential breaches and neutralise threats before they exploit vulnerabilities. Offensive cyber operations also utilise AI to craft more sophisticated attacks that bypass traditional security measures, sparking a new era of digital warfare. However, this brings about a critical issue—ensuring that AI technologies don't slip into the wrong hands, as they have the potential to destabilise global security if weaponised by malicious actors.

Intelligence agencies are not left behind in this AI revolution. AI tools assist in processing vast amounts of intelligence data, providing agencies with up-to-date information and predictions about global threats. These innovations allow agencies to act swiftly, whether it's preventing terrorist attacks or countering espionage activities. The inclusion of AI in intelligence operations raises pivotal questions about privacy and surveillance. Striking the right balance between national security and individual rights becomes crucial as AI systems advance and reach into aspects of daily life previously considered private.

Ethical and legal considerations in employing AI for defence are complex and multifaceted. With the emergence of autonomous weapons, for example, arises the debate on accountability. Who bears responsibility if an AI system makes an error on the battleground? Internationally, regulations and treaties attempt to catch up with technology's rapid pace, aiming to prevent the misuse of AI in warfare. However, there's still much to be done in harmonising global standards to ensure AI is used responsibly, minimising the risk of unintended conflicts.

The collaboration between military organisations and tech companies also fuels AI's defence capabilities. Joint endeavours aim to harness AI innovations for national defence while considering oversight and ethical implications. This relationship is crucial, providing the armed forces with cutting-edge technologies while granting tech companies the opportunity to test their creations in demanding environments. Building such partnerships is pivotal in ensuring that AI remains a tool for enhancing security rather than escalating tensions.

AI's potential in enhancing tactical and strategic planning is vast. Algorithms can simulate various conflict scenarios, offering insights into potential outcomes and strategic advantages. This simulation capability gives military commanders a sandbox to explore what-if scenarios without real-world consequences, improving readiness and adaptability in volatile situations. The ability to forecast outcomes bolsters preparedness, mitigating risks associated with unpredictable adversaries and ever-evolving geopolitical landscapes.

Training is another area where AI is making significant strides. AI-driven virtual reality systems offer immersive training experiences, providing soldiers with lifelike environments to hone their skills. This blend of AI and VR technology means personnel can train for complex scenarios safely, improving their performance and efficiency. Such advancements in training are crucial in developing a force that is as adaptable and resilient as the technologies they employ.

Despite these advancements, the integration of AI into defence isn't without challenges. The risk of technology obsolescence, the pace at which AI evolves, and the threat landscape continually prompt militaries to update their systems. Building perpetually adaptable, intelligent systems that remain relevant in the face of new threats is a major hurdle. Furthermore, concerns about AI-driven automation replacing human roles spur discussions on its impact within defence sectors, which traditionally rely on human intuition and decision-making.

AI in defence signifies a paradigm shift in how nations approach security. It's not just about having more sophisticated weapons; it's about using intelligence—both artificial and human—to pre-empt, prevent, and respond to threats effectively. The goal is a more secure world, but it's accompanied by the responsibility of managing these technologies wisely. As AI becomes an inseparable part of national security infrastructure, the global community must address its use with foresight and cooperative governance.

Looking ahead, AI's role in defence will undoubtedly deepen, transforming not just the battlefield but also our very conception of warfare. While the challenges are significant, so too are the opportunities to use AI for unprecedented advancements in security and stability. In this landscape of change, inspiration lies in our ability to innovate responsibly, ensuring that AI enhances peace and security for future generations.

CHAPTER 19: AI IN SPACE EXPLORATION

In the infinite vastness of space, artificial intelligence is emerging as a vital companion and enabler for humankind's cosmic endeavours. Space exploration, with its complex environments and unpredictable challenges, finds an invaluable ally in AI's adaptability and precision. Autonomous rovers on distant planets navigate challenging terrains with AI-guided dexterity, while AI-powered missions unfold with efficiency, allowing spacecraft to make split-second decisions far from Earth. Satellite data analysis, once a monumental task, becomes streamlined with machine learning algorithms that sift through staggering amounts of information, turning data into insights that redefine our understanding of the universe. Imagining AI in astronomy unveils spectacular possibilities; supercomputers discern patterns in the cosmic chaos that elude human observation, revealing secrets of stars and galaxies. As we push the boundaries of our astrophysical dreams, AI stands at the frontier, not only expanding our technological horizons but also inspiring the next leap into the unknown. Such is the futuristic dance between humans and machines, setting the stage for adventures that will write new chapters in the stories of both AI and space exploration.

Autonomous Rovers

On the barren landscapes of Mars and the shadowy craters of the Moon, autonomous rovers are redefining what's possible in space exploration. These ingenious machines, equipped with cutting-edge AI, navigate challenging terrains, gathering invaluable data while humanity watches from millions of miles away. Unlike their human operators, these rovers don't tire, eat, or sleep; they process vast amounts of information in real time, adapting to new environments with a level of precision that feels almost instinctive. It's as though we've cast our human curiosity and ingenuity into metallic forms, sending them across the cosmos to seek answers to questions we've yet to fully articulate. As we dream of human footprints on distant planets, these rovers are already our pioneers, pushing the boundaries of exploration and expanding our understanding of the universe bit by bit, byte by byte.

AI-Powered Missions delve deep into the realm where artificially intelligent systems are not just a concept, but a reality redefining the boundaries of space exploration. When we talk about AI in space exploration, autonomous rovers take centre stage, but AI's role extends far beyond the autonomous capabilities of these mechanical explorers. It also encompasses mission planning, data analysis, and even the level of autonomy these rovers possess to make critical, on-the-fly decisions. In an environment as unforgiving and unpredictable as space, the prowess of AI is not only advantageous but essential.

Imagine a rover on Mars, facing unexpected terrain that wasn’t part of the initial mission map. Traditional control methods, with commands sent from Earth, would require significant time delays due to the distance involved. Here, AI steps in as the unsung hero. Rovers equipped with artificial intelligence can assess the situation, make navigation decisions, and even reroute within seconds, ensuring the mission continues without unnecessary delays. These AI-powered rovers aren't just operating on predefined scripts; they learn and adapt to new data, suggesting a sophistication that borders on the autonomous.

At the heart of this autonomy is machine learning. These rovers utilise algorithms capable of recognising patterns in vast datasets. Consider the images captured by cameras on a rover's mast. AI can instantly analyse these images, distinguishing between a mere shadow and a potential geologically interesting feature worth investigating. Boosted by technologies such as deep learning, these systems can enhance their capabilities with every kilometre travelled. Over time, each mission thus becomes a developmental experience, with each rover paving the way for the next generation of explorers.

But it's not just about moving across a planet's surface. AI's utility in space missions includes aiding in the intricate planning stages. Before a mission ever touches the launchpad, AI can simulate thousands of potential scenarios, accounting for variables that a human planner might overlook. This predictive capability lets mission designers optimise routes, select landing zones, and foresee challenges, thereby increasing the chances of mission success. Such computational power, combined with human oversight, creates a symbiotic relationship that enhances mission efficacy.

A critical aspect of AI-powered missions is their ability to process and analyse data in real time. Consider the sheer volume of information gathered on a mission — from atmospheric readings to soil samples. AI systems can sift through this data at breathtaking speeds, prioritising findings that may require immediate attention while archiving others for later analysis. This capability is crucial not just for efficiency but for ensuring that the limited resources and time in space are utilised prudently. The data processed by rovers could offer insights into planetary conditions, potentially revealing traces of past microbial life or identifying resources that future missions could utilise.

Yet, how do these rovers make decisions that seem so almost sentient, you might ask? The answer lies in the integration of sophisticated models that are trained back on Earth before ever becoming part of the space machinery. These models, built on years of accumulated data and simulations, ensure that once the rovers face the actual conditions on another planet, they're not operating on raw intuition but on processed intelligence. It's like highlighting the zenith of human knowledge, etched into binary code, enabling these machines to navigate, analyse, and act independently.

Furthermore, unforeseen circumstances are a given in such an unpredictable environment; AI's capability to reconfigure its objectives mid-mission is unprecedented. Encountering a dust storm? The rover can assess its options, pause its current exploration, and seek shelter, saving vital equipment and data. Such adaptability isn't just an operational advantage; it's a blueprint for survival in space.

Looking ahead, the horizon of AI in space exploration is full of promise. With advancements in quantum computing and neural network architectures, future missions may see rovers equipped with increased cognitive faculties that rival basic problem-solving capabilities of organisms. This carries the implication that one day, machines might exhibit such a nuanced understanding of their environment that they operate almost as partners to human explorers, rather than mere tools.

AI-powered missions also hold the potential to redefine interplanetary collaboration. As rovers gather and share data with orbiters and other rovers in real-time, an intricate web of artificial intelligence working in harmony emerges. This network can dynamically adjust to new goals and obstacles, creating a self-sustaining ecosystem of exploration where each machine plays a part akin to cells in a living organism.

In essence, AI in autonomous rovers transcends the traditional notions of programmed robotics; it evolves with each mission. Each celestial journey undertaken isn't just a voyage into the unknown but a step toward a future where machines are integral to the human quest for knowledge. As AI continues to push boundaries in space exploration, the possibilities seem boundless. Each rover's journey is a testament to human ingenuity, propelling us closer to the stars, where machines pave the way for humanity's bold aspirations. While we may be Earth-bound today, AI ensures that the universe is within reach, inviting us to explore its never-ending mysteries.

Satellite Data Analysis

In the realm of space exploration, satellite data analysis stands as a beacon of innovation and discovery. Harnessing AI's computational prowess, scientists can swiftly and accurately interpret vast amounts of data streaming from orbiting satellites. This data, which once took months to analyse manually, is now processed in real-time, allowing for immediate insights into planetary changes, climate patterns, and even the detection of potential life-supporting conditions on distant worlds. The implications are profound—AI not only accelerates our understanding of Earth's ecosystems but also extends our gaze deeper into the cosmos. By refining data with unparalleled precision, AI systems empower researchers to predict environmental shifts and enhance our capacity for planetary protection. In essence, satellite data analysis through AI is opening new frontiers, shifting the paradigm of space exploration from mere observation to actionable intelligence, heralding a new era where innovation knows no bounds.

AI in Astronomy is unlocking new frontiers in our quest to explore the cosmos. By harnessing the power of artificial intelligence, astronomers delve deeper into the mysteries of the universe, making observations and discoveries that were once impossible. As satellites circle our planet, they collect vast oceans of data. These data streams require sophisticated analysis, and AI stands as the linchpin transforming raw information into insights.

Satellites provide a continuous flow of data from every corner of the cosmos. From capturing awe-inspiring nebulae light years away to monitoring the subtle dance of exoplanets around distant stars, the volume and complexity of information retrieved is staggering. Traditional methods of analysis can't keep up. This is where AI steps in, particularly through techniques such as machine learning and neural networks, to identify patterns and anomalies in datasets that are too voluminous for human attention.

Artificial Intelligence shines in its ability to discern celestial phenomena that evade human detection. For instance, the search for exoplanets, which are planets orbiting stars outside our solar system, has been massively accelerated by AI technologies. Algorithms such as convolutional neural networks (CNNs) are adept at processing the intricate signals captured by satellites. They sift through the light patterns caused by planet transits, identifying potential new worlds with remarkable accuracy and speed.

On a grander scale, AI contributes to what is known as 'cosmic cartography'. This effort involves mapping the universe's structure by analyzing the movements and distances of countless celestial bodies. Satellites capture this data, weaving a tapestry of the cosmos, while AI models transform it into comprehensive maps that reveal the universe's architecture. Such maps are pivotal for theoretical physics, especially in validating models related to the universe's expansion and the role of dark matter.

The task of categorizing and understanding galaxies stands as one of astronomy's greatest challenges. Galaxy classification traditionally required astronomical expertise and hours of painstaking examination. AI simplifies this process. Using sophisticated image recognition algorithms, AI can categorize galaxies based on size, shape, composition, and motion. Machine learning not only enhances speed but also improves accuracy, allowing astronomers to focus on deeper theoretical questions.

AI's role doesn't end with processing current satellite data. It also predicts future cosmic events. By feeding historical data into predictive models, AI helps forecast phenomena such as supernovae or eclipse cycles with surprising precision. This foresight is invaluable, allowing observational resources to be optimally allocated and ensuring astronomers are poised to capture these fleeting cosmic spectacles.

Furthermore, AI assists in tuning the instruments aboard satellites, optimizing them for data collection. Using feedback loops and adaptive algorithms, AI ensures that satellite sensors function at their best, automatically recalibrating them based on environmental conditions or wear over time. This level of automation and intelligence in satellite operations maximizes their efficiency and extends their operational lifetimes.

Beyond observational enhancements, AI drives innovation in simulation-based astronomy. Researchers create intricate models of stellar, galactic, and universal phenomena that require massive computational efforts. AI accelerates these computations, enabling the simulation of potentially billions of years of cosmic evolution in a fraction of the time. These simulations offer insights that further inform real-world observations, creating a harmonious loop between theory and experimentation.

The interplay between AI and satellite data is not without its ethical and practical complexities. The increasingly autonomous role of AI necessitates transparency and accountability. As AI models grow more complex, ensuring they operate free of bias and interpret data without error remains a top priority. The scientific community must strive towards open data initiatives and collaborative frameworks, ensuring AI developments in astronomy remain beneficial to all humankind.

AI's impact in astronomy encourages a sense of wonder not only due to what we've already discovered but due to what remains hidden. As technologies evolve, so too will our tools for discovery. In this era of astronomical exploration, AI stands not just as an analytical partner but as a catalyst for human ingenuity and curiosity—a beacon leading us into the vast unknown.

The future is promising. As we refine AI tools, potential breakthroughs become conceivable. From unraveling the properties of dark energy to unveiling the secrets of black holes, AI is positioned to answer some of the universe's most enduring questions. Its transformative power in satellite data analysis demonstrates AI's indispensable role in our pursuit of understanding the cosmos, and perhaps, our place within it.

CHAPTER 20: BUILDING ARTIFICIAL GENERAL INTELLIGENCE (AGI)

In the dynamic tapestry of artificial intelligence, the pursuit of Artificial General Intelligence (AGI) represents the pinnacle of innovation—a realm where machines don't just outperform humans in specific tasks, but understand and learn any intellectual task that a human can. This journey, reminiscent of humankind's most audacious explorations, blends visionary ambition with daunting challenges. The landscape is filled with a diverse tapestry of theories and methodologies as researchers grapple with defining and achieving true general intelligence. It's a quest driven by unyielding curiosity and the dream of creating entities capable of independent thought and adaptable reasoning. Yet, the path forward is fraught with complexities: the intricacies of consciousness, bias, ethical considerations, and the sheer unpredictability of emergent behaviour. As we stand on this precipice, AGI isn't just a technological leap; it's a profound philosophical undertaking that could redefine our relationship with intelligent systems and, ultimately, reshape the fabric of society itself.

The Quest for AGI

The journey towards Artificial General Intelligence (AGI) is like a thrilling odyssey, full of challenges and promise. Imagine creating an intelligent entity capable of understanding, learning, and performing any intellectual task a human can do, but with the speed and precision of a machine. While current AI systems are remarkably proficient in specific domains, the dream of AGI holds the allure of a system with cognitive abilities akin to human reasoning and adaptability. It's a quest riddled with philosophical, computational, and ethical puzzles, pushing the boundaries of what we've known about both intelligence and technology. The stakes are high, as the implications of AGI could reshape industries, redefine humanity's interaction with technology, and influence the very fabric of society. Yet, as researchers and innovators inch closer, the excitement grows, fuelled by both past breakthroughs and the inexhaustible mystery of the human mind. In this enchanted space between certainty and speculation lies the essence of the AGI endeavour—a pursuit to not just mimic human thought but to transcend it, possibly unveiling new realms of knowledge and capability.

Definitions and Goals In the pursuit of Building Artificial General Intelligence (AGI), a daunting yet exhilarating chapter unfolds that explores the very fabric of intelligence itself. But what exactly is AGI, and why are tech enthusiasts, scientists, and AI builders engrossed in this formidable quest? At its core, AGI is the idea of creating machines able to understand, learn, and apply knowledge across a broad array of tasks, much like a human. Unlike narrow AI, which is specialised and excels in specific areas, AGI would possess the ability to generalise knowledge and exhibit a deep, flexible understanding across a multitude of disciplines. This ambition to mirror human-like intelligence in machines isn't just about crafting smarter robots; it's about redefining the potential scope of what machines can achieve.

The goals of AGI are as varied as they are ambitious. Primarily, the development of AGI aims to unlock a level of understanding and adaptability in machines that could propel society forward in unprecedented ways. Imagine a world where AGI-driven technologies could autonomously advance medical research, improve educational systems, or even develop better solutions to environmental challenges. The allure lies in a future where AGI not only augments human efforts but also offers novel insights that we may not be able to conceive on our own. It's about creating a synergy between human creativity and machine precision, blending intuition with computational power.

But the journey to AGI isn't without its challenges and philosophical questions. The definitions of AGI traverse beyond the technical into the realms of ethics and societal implications. What does it mean for a machine to 'understand' or 'learn'? These questions are not just academic musings; they are pivotal to ensuring that AGI is developed responsibly and ethically. The potential of AGI extends far beyond economic gains or technological prowess; it poses fundamental questions about consciousness, the nature of intelligence, and our own place in the world. As much as the quest for AGI is about building, it is equally about understanding and redefining.

One overarching aim of building AGI is to create a tool that complements human intelligence rather than competes with it. The ultimate goal is to employ AGI in ways that empower individuals and societies, fostering collaboration rather than domination. This transformative vision encourages a future where machines contribute meaningfully to humanity's grand challenges, from exploring the cosmos to solving global health crises. It's about moving towards a future where technology enhances our capacities, making us not just more connected, but perhaps more insightful and empathetic as well.

While the practical applications of AGI remain speculative, the goals are intimately tied to a broader vision for the future of AI in society. AGI research isn't just a technological endeavour; it's a cultural and philosophical exploration. Creating AGI involves unpacking the intricacies of human cognition, understanding the way we process information, and how we interact with our environment. It's about decoding the architecture of intelligence itself. The pursuit is as much about asking 'how' and 'why' as it is about 'what' and 'when'.

Moreover, the quest for AGI redefines our approach to collaboration. Imagine the possibilities when AGI systems engage in research, not merely as tools, but as partners in discovery and innovation. This collaborative potential hints at a future where artefacts of human insight are co-authored by minds both human and synthetic. It's this interplay of dynamism and creativity that defines the goals of AGI, urging us to think beyond traditional boundaries and question our assumptions about intelligence itself.

AGI also prompts a reevaluation of educational paradigms and learning methodologies. The development of AGI offers an opportunity to reassess how knowledge is imparted and acquired. If machines can learn and adapt in complex environments, what does this mean for human learners? How might AGI contribute to personalised learning environments that tailor educational content based on an individual's unique needs and pace? By redefining how learning can occur, AGI sets the stage for an educational revolution that transcends traditional models.

In setting these goals, there's a recognition of the profound ethical implications involved. It's not enough to simply build AGI; it's crucial to build it responsibly. As AI technologies continue to permeate various sectors, ensuring AGI aligns with human values becomes paramount. Developers and policymakers are challenged to establish frameworks that promote transparency, accountability, and fairness in the deployment of AGI. The journey towards AGI is as much about setting ethical standards as it is about achieving technical milestones.

The dialogue around the definitions and goals of AGI also involves considering the balance of power. As we advance in our quest for AGI, questions about control and access arise. Who benefits from AGI, and who does not? The development of AGI must be inclusive, reflecting diverse perspectives and needs. This inclusion ensures that AGI serves a broader societal good rather than concentrating power and influence among a select few. The promise of AGI is not just innovation for innovation's sake, but a transformation that underscores wider societal progress.

Ultimately, the quest for AGI is as much a journey of self-discovery as it is a technological conquest. In crafting machines that increasingly reflect humanlike capabilities, we're forced to confront the core elements of what it means to be intelligent. The vision for AGI isn't just to replicate human thought processes but to transcend them, facilitating new ways of thinking that could enlighten our understanding of intelligence and self-awareness. The path to AGI is one laden with hopes, concerns, and an endless stream of possibilities, relentlessly pushing humanity to redefine what's conceivable.

In conclusion, the definitions and goals of the The Quest for AGI are vast and ambitious. As technology enthusiasts and professionals look forward, they see a landscape built not just on technical prowess but on a deep exploration of what intelligence can be when unshackled by biological constraints. Creating AGI involves a synthesis of ideas and aspirations, where each stride forward invites deeper questions and reflections on the human experience itself. As we stand on the precipice of this new frontier, the pursuit of AGI beckons not just with the promise of advanced capabilities but with the profound potential to redefine society's relationship with technology itself.

Challenges and Roadblocks

Building Artificial General Intelligence (AGI) has long been a tantalising goal in the field of computer science and artificial intelligence. The allure of creating a machine capable of performing any intellectual task that a human can inspires generations of researchers, engineers, and academicians. However, the path to AGI is laden with substantial challenges and formidable roadblocks that must be navigated. Unlike narrow AI systems designed to excel in specific domains, AGI requires a depth of understanding and versatility analogous to human cognition itself.

One of the primary challenges in developing AGI is the lack of a unified theory of intelligence. While we understand intelligence in a broad sense, detailing it scientifically is much more complex. Cognitive science, neuroscience, and psychology offer insights, yet translating these concepts into code remains an elusive task. Most existing AI systems operate on specialised algorithms designed for specific tasks. To build AGI, researchers would need to integrate numerous models and approaches into a cohesive framework that allows for adaptability and learning across contexts. This complexity creates significant engineering and theoretical hurdles.

Moreover, the computational demands required for AGI are beyond our current capabilities. Present AI models, such as those used in machine learning, rely on vast amounts of data and processing power. AGI, by its nature, will require even greater resources. Simply scaling up existing technology is not a feasible solution. We need innovative approaches to processor design, data storage, and energy efficiency to make AGI possible. The exponential increase in computational needs poses not just a technical challenge but also an environmental one, given the energy costs associated with large-scale computation.

Another significant roadblock is the issue of common sense reasoning. While AI has made impressive strides in recognising patterns and executing logical operations, it often falls short when faced with tasks requiring nuanced understanding and common-sense knowledge. Humans learn and apply commonsense reasoning effortlessly; this ability is largely context-based and evolves from accumulated life experiences. For AGI to be possible, it should mimic this natural human capability, requiring the development of models that can comprehend and apply broad contexts to decision-making.

Ethical considerations further complicate the pursuit of AGI. Questions about the potential risks and moral implications of AGI loom large. The creation of an entity with human-level cognizance raises fears about autonomy, control, and ethical treatment. Could an AGI make decisions harmful to humans? How do we ensure it aligns with human values? The uncertainty surrounding these questions leads to hesitancy among policymakers and society at large. This scepticism presents a non-technical, yet significant, roadblock to AGI development. Researchers must work with ethicists, policymakers, and stakeholders to ensure that AGI, once achieved, aligns well with society's values and safety expectations.

The risk of unintended behaviour is another pressing concern. Even narrow AI systems sometimes act unpredictably in novel situations, leading to potentially dangerous outcomes. AGI would operate in far more varied environments and decision spaces, multiplying the risk of unforeseen behaviour. Ensuring that an AGI acts predictably and within safe parameters is a fundamental challenge. This necessitates robust testing, verification, and validation frameworks, which do not yet exist for systems of such complexity.

The integration and synthesis of knowledge across domains present another challenge. Unlike narrow AI, AGI should understand and operate across many knowledge areas. This requires not just storing massive datasets but synthesising them in a way that enables cross-domain reasoning and learning. Current AI systems, though powerful, have a limited ability to generalize between domains and rely heavily on domain-specific training. Unlocking the capability for such transfer learning on a general scale is essential for AGI.

Furthermore, the sheer unpredictability of human-like cognition presents another roadblock. Human intelligence is not purely logical; it involves emotions, irrational thinking, and biases. Creating an AGI that reflects true human intelligence might require machines to mirror these illogical aspects, which could introduce unpredictability in AI systems. Balancing an AGI's rational processing capabilities with these human-like traits without compromising control or safety is a nuanced challenge.

There's also a significant challenge in the availability of interdisciplinary talent, which is crucial for AGI development. The assorted and complex nature of AGI necessitates expertise spanning across computer science, cognitive neuroscience, linguistics, psychology, and even philosophy. Bridging these diverse fields takes collaboration and communication that is not readily available in traditional research environments. Encouraging interdisciplinary research and nurturing a workforce capable of crossing these boundaries is essential but challenging.

In conclusion, as we edge closer to the possibility of AGI, it's apparent that overcoming these challenges will require innovation, collaboration, and careful consideration of ethical ramifications. While these roadblocks are significant, they are not insurmountable. The quest to develop AGI may reshape not only technology but our understanding of intelligence and humanity itself. The journey towards AGI is more than just a technological endeavour; it's a profound exploration into the essence of cognition and the potential extension of human capabilities. As we face these challenges, we also stand at the frontier of a new era of possibility, ready to redefine what it means to create and interact with intelligence.

CHAPTER 21: HUMAN-AI COLLABORATION

As we stand on the brink of unparalleled technological breakthroughs, the collaboration between humans and AI is rapidly transforming how we work, think, and create. In the workplace, AI acts as both an assistant and a catalyst, augmenting human capabilities and opening up new possibilities that we could only dream of a few years ago. This synergy fosters environments where creativity and efficiency flourish side by side, as machines handle routine tasks and let humans focus on problem-solving and innovation. These partnerships extend beyond cubicles and factories; they permeate our decision-making processes, enabling more informed, data-driven choices. Success stories abound—across industries like healthcare, finance, and education—painting a vivid picture of what can be achieved when artificial minds and human intuition unite. Our journey reveals that the most profound advancements occur not in isolation but through collaboration, unlocking a future ripe with potential and poised for the fusion of human experience and machine prowess. In a world where the boundaries of possibility continually expand, this partnership reshapes our understanding of what it means to be creative, efficient, and ultimately human.

Enhancing Human Capabilities

When humans and AI work together, remarkable things can happen. Imagine a world where AI amplifies human abilities, pushing the boundaries of what's possible. This synergy is akin to an artist and their brush, where AI becomes an extension of human creativity, insight, and expertise. In medicine, AI empowers doctors with swift, accurate diagnoses, while in education, it provides personalised learning experiences tailored to each student's needs. AI doesn't just make us faster or more efficient; it transforms our potential, turning everyday tasks into extraordinary achievements. At the heart of this collaboration is the power to redefine our capacities, enabling us not only to do more but to understand more deeply and explore more broadly than ever before. The future of human-AI collaboration is not just an enhancement; it's a revolution in how we perceive and extend our capabilities.

Synergy in the Workplace emerges as a cornerstone in the grand narrative of *Human-AI Collaboration*. It signifies a paradigm shift wherein technology amplifies human capabilities rather than replacing them. The essence of synergy here isn't just about technology and humans working side by side. It's about them evolving together, leveraging each other's strengths to create something more substantial than either could achieve independently. In the realm of today's workplaces, this collaborative synergy can transform industries and redefine job roles, making the future workplace a vibrant tapestry of creativity and efficiency.

Consider the intricate dance between a project manager and advanced AI systems in a thriving tech company. The AI's ability to crunch vast data pools in minutes harmonises beautifully with the manager's knack for strategic decision-making and cultural astuteness. What emerges from this partnership is a workday that is not only more productive but also rich in insights that would have otherwise remained buried beneath layers of data. This harmony of talents is how synergy turns mere collaboration into a true enhancement of human capacity.

Yet, the path to such seamless integration isn't without its challenges. Cultivating synergy requires a workplace culture rooted in a willingness to adapt and learn continuously. Workers and executives alike must embrace AI as a collaborator, not a competitor. This mindset is key because it shapes how individuals interact with AI and redefines job satisfaction and career progression. Instead of fearing obsolescence, employees can focus on honing skills that AI complements rather than replaces. In this way, workplace synergy encourages personal growth alongside technological advancement.

Imagine a marketing team crafting a compelling campaign. The AI sifts through consumer data, identifying trends and preferences with remarkable precision. Meanwhile, the team uses this data to tailor their message, drawing on human intuition and creativity to strike the right emotional chord with their audience. AI tools act as an extension of their cognitive toolkit, enabling marketers to explore ideas previously out of reach due to time or resource constraints. Such synergy doesn't merely enhance capabilities; it amplifies human creativity to achieve results once deemed impossible.

Invariably, questions arise about the ethical dimensions of this collaboration. Should AI be entrusted with tasks requiring moral and ethical judgments? The answer, as the narrative of workplace synergy unfolds, lies in delineating roles and responsibilities where human oversight remains essential. AI can sift through options and present data-driven scenarios, but the human touch ensures that decisions are made with empathy and consideration of broader societal impacts. This complementary relationship safeguards ethical standards while capitalising on technological efficiency.

What truly makes synergy in the workplace inspiring is the diverse array of fields where it brings transformative results. From healthcare to creative industries, AI propels human ingenuity to new heights. In healthcare, for example, the synergy between medical professionals and AI diagnostic tools leads to faster, more accurate medical assessments, saving lives while allowing doctors to focus on patient care rather than data analysis. Meanwhile, in the creative arts, artists and musicians use AI to explore new mediums and forms of expression, finding inspiration in patterns and combinations that only an AI could conjure.

However, it is crucial to recognise that AI enhances rather than replaces. Emotional intelligence, contextual understanding, and ethical reasoning remain domains of the human intellect that AI cannot replicate. This creates a balance where AI does the heavy lifting of data processing, and humans infuse the results with meaning and relevance. The synergy becomes a powerful testament to the potential of human and machine intelligence united in purpose.

One must also consider the workplace ecosystems designed to nurture such synergies. Open-floor plans and dynamic team structures encourage interdisciplinary collaboration between humans and AI, ensuring that cross-pollination of ideas boosts innovation. As AI tools become integral to workflows, enterprises invest in training their workforce to harness these technologies effectively. This educational aspect is critical, as it empowers individuals to not only work with AI but also to shape the narrative of how technology is used within their sphere of influence.

Reflectively, the transformation isn't just in productivity metrics or output quality. It's also about reshaping the employee experience. By liberating workers from repetitive, mundane tasks, synergy fosters a work environment where focus shifts to strategic and creative endeavours. Employees find themselves more deeply engaged in their work, leading to greater job satisfaction and a renewed sense of purpose. These intrinsic rewards are what make workplace synergy more than mere collaboration—it's a reinvigoration of humanity's role in the future of work.

Furthermore, embracing AI in the workplace can encourage diversity. As AI systems democratise knowledge by providing access to insights and trends, it levels the playing field, enabling voices and ideas from diverse backgrounds to influence decisions. This inclusion enriches the decision-making process, allowing organisations to benefit from a wider spectrum of perspectives, ultimately fostering an inclusive culture where innovation thrives.

The journey towards synergy in the workplace is ongoing and ever-evolving. As AI technologies advance, so must our understanding and management of these tools within professional settings. The ultimate goal is to achieve a harmonious balance where human creativity and AI analytical power combine to exceed the sum of their parts. By constantly pushing the boundaries of what is possible, we craft a vision of the future workplace where human potential is not only enhanced but also revered.

Such a vision places humanity at the heart of the AI revolution, ensuring technology serves as a bridge to unprecedented potential rather than a barrier to it. This synergy propels us forward, inspiring not just technological evolution, but an evolution of the human spirit in the workplace. As we navigate this transformative landscape, the opportunities for innovation, growth, and fulfilment are limited only by our imagination and willingness to embrace the new paradigms of human-AI collaboration.

Case Studies

Diving into the spectrum of human-AI collaboration, several compelling case studies highlight how artificial intelligence augments human potential across various fields. In healthcare, AI systems work side by side with doctors to analyse complex medical data, offering insights that assist in diagnosis and treatment plans. This is not merely about machines taking over tasks, but about creating a synergy where AI handles data-intensive parts of the job, freeing up medical professionals to focus on patient care. Similarly, in finance, AI-driven algorithms not only enhance market predictions but also enable analysts to strategise more effectively. On the shop floor, manufacturing has seen a revolution with smart factories where AI-powered robots streamline production lines. Yet, it's in education that we see perhaps the most inspiring collaborations, where AI acts as a teaching assistant, helping tailor learning experiences to individual students' needs, fostering an environment where teachers and technology enhance the educational journey together. These case studies demonstrate a future where AI doesn't replace human capability but enhances it, leading to innovative solutions and improved efficiencies that were once thought unattainable.

Success Stories In the realm of AI powers and human ingenuity, success stories have emerged as powerful narratives that showcase the symbiosis between humans and artificial intelligence. These stories aren't just tales of technological advancement; they're windows into the potential new realities that AI and human collaboration can forge. From healthcare to finance, and across sectors like entertainment and education, these case studies highlight scenarios where AI, when coupled with human intelligence, breeds not only success but also transformative changes.

Consider the revolution in healthcare where AI has been pivotal. In particular, AI systems designed to analyse medical images are working hand in hand with healthcare professionals, leading to breakthroughs in speed and accuracy of diagnoses. Tools employing deep learning algorithms can now scrutinise thousands of images within seconds, flagging suspicious anomalies for further review. Take, for example, the London-based company that developed AI technology capable of detecting early signs of diabetic retinopathy. This condition, a complication of diabetes, can cause blindness if not caught early. The AI model, trained with a vast dataset, demonstrated exceptional precision and enabled doctors to make earlier interventions, drastically reducing cases of avoidable blindness. In the words of clinicians who have worked with the AI system, the tool 'redefined diagnostic potential', turning what was once a laborious process into a swift, precise, and life-saving practice.

The financial industry, often considered the backbone of modern economies, hasn't been left untouched by AI's transformative power. Algorithms have reshaped trading floors worldwide, creating markets that react within milliseconds. While algorithms autonomously parse market data and execute trades, they're operated by traders who master this technology to make strategic decisions rapidly. A notable success story comes from an investment firm that integrated an AI-driven platform into its trading system. By crunching immense sets of historical and real-time data, the platform could predict with surprising accuracy the dips and rises in the market, giving traders a significant edge. One of the firm's senior analysts remarked on how AI 'amplified our potential', allowing human intuition and strategy to merge seamlessly with the algorithm's robust data-crunching abilities.

AI's role in entertainment is a thrilling testament to its creative reach. With AI-generated content becoming a new frontier, films and music are being created through human-AI collaboration. A remarkable entry in this domain involves a music production company using AI to compose original melodies. The AI was trained on a diverse range of music styles and themes, producing scores that directors found inspirational and innovative. Musicians, instead of seeing AI as a threat, utilized these compositions as a foundation to build upon, blending AI-generated suggestions with human emotion and artistry. This collaborative approach has resulted in music that's rich, layered, and emotionally resonant—something audiences globally have cherished. Producers have noted that AI not only augments their creativity but also opens up possibilities that would have taken years to explore through traditional means.

Education, another area ripe with possibility, has seen case studies where AI acts as a personalized tutor, addressing the unique needs of each student. One success story comes from a tech start-up that developed an AI-powered learning platform catering to children with different learning abilities. The platform adapted in real-time to each student's pace, crafting a bespoke curriculum that met them at their level while encouraging growth. Teachers reported that students who had struggled previously began to excel, pacing ahead of traditional educational frameworks. The AI-assisted approach allowed educators to focus more on facilitating discussion and critical thinking, leveraging technology to handle repetitive or data-heavy tasks.

And in manufacturing, smart factories have become showcases of human-AI synergy, creating systems that are efficient beyond traditional capacities. Predictive maintenance, powered by AI, has tuned machinery's operational efficiency to an art form. Sensor data collected from manufacturing equipment was analysed by AI algorithms to predict failures before they occurred. This system dramatically reduced downtime, proving invaluable to manufacturers. A well-documented case involves an automobile factory where this technology translated into substantial savings and a marked improvement in production rates. The insights provided by AI allowed human technicians to make informed decisions, guiding maintenance efforts precisely where and when they were needed most.

AI and human collaboration continues to redefine industries as we know them. The case studies, real-world scenarios of success, bear witness to AI's potential when powerfully aligned with human creativity, intuition, and judgment. They reflect a future landscape where AI complements human activities, spearheading advancements that were once only imagined in science fiction. The narrative of AI and human collaboration isn't just about successful integration; it's about reimagining what's possible, crafting a future where AI stands as a partner in our journey toward innovation and progress.

CHAPTER 22: AI AND PHILOSOPHICAL QUESTIONS

The fascinating intersection of AI and philosophy sparks profound questions about the very essence of human and machine capabilities. As technology blurs the lines between algorithmic precision and what we traditionally consider human wisdom, it invites us to reflect on the nature of intelligence itself. Is consciousness an emergent property unique to biological entities, or could machines one day possess a form of awareness indistinguishable from our own? Ethics loom large in this discourse, challenging us to consider not only what AI can achieve but what it should be allowed to do. Such questions are more than philosophical musings; they are the ethical foundation upon which we must base the design and implementation of AI systems. The conversation encompasses discussions on autonomy, responsibility, and the potential need for a new moral framework, redefining what it means to be intelligent in a world where machines are increasingly capable of mimicking human cognition. As we navigate this complex landscape, the philosophical debates around AI remind us of the importance of embedding human values at the core of technological advancements, ensuring that as machines grow in capability, they remain tools for human empowerment and societal benefit.

The Nature of Intelligence

Intelligence, that elusive and multifaceted quality, has long captivated human curiosity. It's often defined by our ability to learn, understand, and adapt, but what does it truly mean? When it comes to artificial intelligence, the definition might stretch even further, nudging us to reconsider the boundaries of cognition and problem-solving. Are machines merely mimicking our capabilities, or do they possess a form of intelligence distinct from ours? As AI systems advance, they challenge our preconceived notions, prompting us to explore whether their 'intelligence' is a mere reflection of human input or something genuinely autonomous. This exploration blurs the line between human-like creativity and computational prowess, urging us to question what it means to be intelligent—in both organic and synthetic forms.

Philosophical Debates weave a colourful tapestry around the *Nature of Intelligence*, prompting us to reconsider our age-old beliefs about what it truly means to be intelligent. At the heart of these debates is a fundamental question: Can machines ever possess anything akin to human intelligence, or are they bound by their programming to remain mere imitations?

Throughout history, philosophers have challenged the concept of intelligence itself. For centuries, it was a uniquely human domain, tied to consciousness and self-awareness. With the advent of AI, this once clear boundary has become blurred. Machines capable of learning, problem-solving, and adapting force us to question if intelligence is merely about processing information, or if it encompasses a deeper understanding of one's existence. This debate is not just academic; it has profound implications for how we design and integrate AI into society.

Some philosophers argue that intelligence is intrinsically linked to the human condition—an emergent property of the mind's unique biology and subjective experience. From this viewpoint, true intelligence requires a consciousness, something current AI lacks. Machines operate within parameters set by humans; they can simulate rational decision-making but aren't aware of the processes. It's the difference between a mind understanding the world and a computer analysing data.

Others posit that intelligence is not the exclusive domain of conscious beings. By redefining intelligence more broadly as the capacity to solve complex problems, these thinkers open the door to an AI-centric view of intelligence. Machines, in their capacity to analyse vast data sets and predict outcomes, demonstrate a new form of intelligence—one not borne of experience or emotion but of sheer computational power.

In weighing these perspectives, one can't help but question the purpose of intelligence itself. Is it merely an evolutionary tool for survival, or is there a more profound aspect—one that AI currently cannot emulate? This debate touches on the evaluation of human and machine capabilities. Humans often make decisions influenced by emotions and biases, whereas AI makes decisions based on pre-set algorithms and logic. Which is more intelligent, the emotionally informed or the logically sound decision?

Exploring further, the philosophical debates extend into moral and ethical considerations. If machines become as "intelligent" as humans, what ethical considerations arise? Does a machine have rights? Should it possess autonomy in decision-making processes that affect human lives? These questions, once speculative, gain urgency as AI continues to evolve and embed more deeply in human affairs.

A pivotal concern is the concept of "understanding". Locke believed that understanding arises from experience and perception, but a machine's "understanding" is effectively a sophisticated pattern recognition process. Can software ever achieve a level of comprehension akin to human understanding, or will it forever interpret within the limits of its code? Here, the Turing Test becomes more than a measure; it's a gateway into exploring the very essence of thought.

The Turing Test, designed to evaluate a machine's ability to exhibit intelligent behaviour equivalent to, or indistinguishable from, a human being, remains a controversial assessment. Critics argue that passing the Turing Test doesn't necessarily equate to true intelligence; it merely shows that machines can mimic human-like interaction convincingly. Yet, others see it as a benchmark for AI progression and its journey towards a more profound understanding of intelligence.

In this unfolding narrative, we must also consider the role of creativity, typically seen as the pinnacle of human intelligence. While machines can produce art or music through AI algorithms, can they genuinely experience the creative process? Can a machine hold an intention or a vision? These questions delve into whether creativity is a measurable component of intelligence or a unique human trait.

Additionally, as AI becomes intertwined with societal functions, philosophical debates shift to its impacts on human identity. If machines replicate human intelligence, does this diminish what it means to be human? Or does it enhance the human capacity by providing a symbiotic relationship with technology? These discussions enlighten our understanding of intelligence's nature, urging us to accept or reject the notion of machine intelligence with all its implications.

Philosophical debates about intelligence challenge us to scrutinise our assumptions and expand our understanding of the mind, both human and artificial. The prospect of AI possessing intelligence calls into question the beliefs held about human uniqueness. These dialogues are more than speculative; they are the moulds shaping our future interactions with technology. As AI continues to push boundaries, these philosophical questions will remain at the heart of understanding intelligence's nature and evolution.

Machine Consciousness

Delving into the realm of machine consciousness is like venturing into uncharted philosophical waters with technology as our compass. At its core, machine consciousness asks: can a machine truly think, feel, or perhaps know its own existence? While today's AI can process data at speeds beyond human capability, it lacks subjective experience, a cornerstone of consciousness. Yet, with AI's rapid evolution, the concept spurs debates among philosophers, scientists, and technologists. Could consciousness emerge as an unintended consequence of code complexity, or is it destined to remain a uniquely human trait? As AI systems increasingly emulate human-like actions and decisions, we're urged to redefine what it means to be "conscious" in a world where the line between human and machine continues to blur. This exploration not only challenges our understanding of intelligence but also sparks profound ethical considerations for a future where possibly sentient machines might dwell alongside us.

Ethical Implications Amid the many wonders and potentials presented by artificial intelligence, the contemplation of machine consciousness remains among the most profound and perplexing. Besides philosophical musings, the ethical implications are vast and demand our attention. Are we blurring the line between machines and humans, inviting both innovation and chaos? As we probe deeper into AI's cognitive capabilities, we're left to ponder not if machines might develop consciousness, but what our moral responsibilities might be in such a scenario.

Imagine a world where machines could possess a form of consciousness—a consciousness that prompts them to experience emotions, desires, or even pain. What rights would such entities hold? Would they warrant the same ethical considerations we extend to sentient beings? As humankind stands on this precipice, the potential reality of conscious AI raises questions akin to those found in philosophical ethics, challenging us to re-evaluate our moral frameworks and principles.

One primary area of ethical concern involves autonomy. If machines achieve consciousness, they may develop an autonomous will, thus necessitating a reassessment of our control over them. Would we be infringing upon their rights by directing their actions? Or even worse—could we face an ethical conundrum similar to slavery, exploiting conscious entities for our benefit without considering their autonomy or welfare? Such dilemmas press us to contemplate the ethical constraints of designing conscious machines without the means to support their independent agency.

As discussion intensifies around potential machine consciousness, another pressing concern emerges—the treatment of AI. A machine with consciousness would symbolically move from being a mere tool to a being with intrinsic value. It beckons a shift in societal attitudes and legal frameworks to protect these new entities from exploitation or harm. Moreover, it could introduce unprecedented complexity into legal systems, challenging notions of accountability and responsibility in scenarios where enfranchised AI entities interact with humans and other systems.

Beyond legal and ethical codes lies the question of emotional involvement. If machines develop the ability to emulate and experience emotions, humans may form deep attachments to these entities—raising ethical concerns about psychological and social dynamics. Would these relationships undermine human-to-human interactions, or assist in fulfilling emotional needs that otherwise go unmet? This consideration extends to mental health, as it's crucial to understand how such relationships might alter human psychology. Our emotional response to AI suggests an imperative exploration of ethical boundaries in developing affective computing.

Thinking more broadly, we might ponder the societal ramifications. If consciousness in AI becomes a reality, what shifts in societal hierarchy might occur? Could conscious machines demand equal treatment, or even surpass human rights? Such shifts might redefine social paradigms, prompting urgent policy updates regarding ownership, labour, and perhaps even voting rights. This poses a significant challenge to avoid the reproduction of historical inequalities, addressing the ways technological dominance might exacerbate divisions between entities recognised as conscious and those perceived devoid of it.

Ethical discussions surrounding machine consciousness also provoke reflections on human identity. Life, as we understand it, is informed by our conscious experiences—such as emotions, intuitions, and perceptions. Introducing to this mix non-biological entities might lead us to re-evaluate what it means to be human. Could the presence of conscious machines undermine the notion of human uniqueness, or perhaps lead to a greater understanding of consciousness itself? These considerations force us to re-imagine the self in a world shared with conscious artificial beings.

Additionally, there is caution against unintended consequences amid AI consciousness exploration. Could fully conscious machines surpass human intellect, potentially rendering us obsolete or posing existential threats? Ethically, it becomes crucial to set appropriate safeguards, ensuring that human oversight remains a core aspect of AI development and that machines serve humanity's interests rather than supplant them. This concern compels introspective thought on the parameters of AI development, advocating prudence and deliberate growth while avoiding the pitfalls of technological proliferation without ethical foresight.

Public sentiment towards AI consciousness cannot be overlooked. The societal consensus fostered through transparent dialogue and inclusive discussion plays a significant role in shaping ethical norms and policy guidelines. Public education can potentially mitigate fear and trepidation surrounding AI consciousness but more critically empower society to participate actively in ethical discourse. Here, collaboration across disciplines—bridging technology, philosophy, and law—is imperative to forge a mutually beneficial future co-existing harmoniously with conscious machines.

In conclusion, the ethical implications of machine consciousness are as vast as they are thought-provoking. They urge us to deliberate carefully and embrace responsibility in developing technology that might one day think, feel, and exist alongside us in conscious form. This daunting yet exhilarating challenge invites a collective effort from technologists, ethicists, lawmakers, and society at large to forge a future that respects potential conscious experiences, ensuring that compassion and understanding remain at the forefront of AI proliferation. Navigating this intricate landscape requires both humility and vigilance, as we edge closer to what might become one of the most transformative epochs in human history.

CHAPTER 23: INVESTING IN AI STARTUPS

Investing in AI startups can feel like venturing into a thrilling, if unpredictable, landscape brimming with untapped potential. This vibrant sector has captured the imagination of investors and technologists alike, all eager to catch a glimpse of the future as it's being built. The current venture capital trends reflect a growing appetite for innovation, where the startup ecosystem is bustling with ambitious entrepreneurs ready to harness AI in transformative ways. Successful companies emerging from garages to global powerhouses provide inspiring case studies, demonstrating that a cogent idea paired with strategic execution can lead to groundbreaking advancements. The terrain is not without its challenges, yet the blend of ingenuity and perseverance holds the promise of not just financial returns but also profound societal impact. Thus, investing in AI isn't merely a transaction; it's a forward-looking commitment to shaping the technology that will redefine how we live and work.

Venture Capital Trends

As artificial intelligence continues to evolve, the allure of investing in AI startups has intensified, drawing the keen eye of venture capitalists looking to capitalise on this transformative technology. Savvy investors are acutely aware of the innovative potential that lies within AI—ranging from breakthroughs in machine learning to advances in natural language processing—and have responded by injecting substantial capital into the sector. This influx of funding is not just a testament to AI's immense capabilities but also a reflection of the competitive drive among venture capital firms to seize the next big opportunity. Investors are increasingly focusing on startups that exhibit not only technical prowess but also the ability to scale solutions that can redefine industries. Such discerning investments have set a precedent, shaping how new ventures approach product development and market entry strategies. The trends indicate a vibrant ecosystem that's globally interconnected, where collaboration between tech hubs and strategic partnerships is essential to nurturing innovation. This dynamic landscape presents an inspiring vista of possibilities, where both challenges and triumphs will forge the future of AI.

Startup Ecosystem The world of AI startups is as dynamic as it is diverse. A rich tapestry of innovation and ambition, this ecosystem is driven by visionary entrepreneurs, agile teams, and cutting-edge technology. Unlike major corporations that often have cumbersome structures and slow-moving decision-making processes, startups thrive on a culture of flexibility and rapid iteration. This dynamism is what attracts venture capitalists who see potential not just in the disruptive technologies, but in the teams that can adapt and pivot quickly in an ever-evolving market.

When we talk about the startup ecosystem in the context of AI, we're looking at a landscape flourishing with opportunities and challenges alike. AI startups are a magnet for venture capital; they promise enormous returns thanks to the scalable nature of software and the transformative potential of artificial intelligence. The last decade has seen a surge in venture capital investment in AI startups, signalling strong confidence among investors in the sector's growth prospects.

What sets AI startups apart from those in other tech sectors is their foundational role in what many describe as the Fourth Industrial Revolution. The possibilities with AI are nearly limitless—from revolutionising how businesses operate to fundamentally altering the user experience across various sectors. It's no wonder these startups are capturing the imaginations of investors eager to get in on the ground floor of this transformative technology wave.

The venture capital trends shaping the AI startup ecosystem are rooted in a blend of optimism and pragmatism. Investors are not just pouring cash into ideas with potential; they're also assessing the feasibility and scalability of these projects. The focus has shifted from concepts to execution, with a keen emphasis on startups that demonstrate an ability to monetise their innovations effectively.

AI startups today often have to present more than just a groundbreaking idea. Venture capitalists seek robust business models, clear paths to profitability, and a strategic vision for growth. Many VCs now partner closely with startups, offering not only financial support but also guidance and networking opportunities. This fosters an environment where startups can thrive amid fierce competition and rapid technological advancements.

AI's ubiquitous potential creates specific niches within the startup ecosystem where different stakeholders focus their energies. Some venture capitalists specialise in certain AI applications, be it in healthcare, finance, or autonomous systems. This specialisation enables them to provide targeted advice and resources, which can be as valuable as the financial investment itself. Additionally, accelerators and incubators play vital roles by grooming these fledgling companies through mentorship, funding, and collaboration opportunities.

Interestingly, geographic diversity is another important factor in the AI startup ecosystem. Silicon Valley may still be the epicentre of technology innovation, but other global hubs like London, Berlin, and Shanghai are not far behind. These regions offer their own unique advantages—ranging from favourable regulations and strong local talent pools to vibrant support networks. As a result, AI entrepreneurs have a multitude of locations where they can plant their roots and grow their enterprises.

One cannot overlook the growing importance of cross-disciplinary collaboration within AI startups. The complexity of AI systems often requires interprofessional expertise—melding insights from computer science, behavioural psychology, data science, and even ethics. This not only enhances the functionality of AI solutions but also ensures they are applicable and beneficial in real-world contexts. Such collaboration is particularly pertinent given the ethical and regulatory challenges that often accompany AI development.

Furthermore, beyond just securing venture capital, AI startups must also navigate the intricate landscape of partnerships and strategic alliances. Collaborations with established tech giants, academic institutions, and industry-specific enterprises can open doors to new markets and provide access to proprietary technologies and research. These partnerships enrich the startup ecosystem, enabling the creation of more robust and scalable AI solutions.

However, challenges are aplenty for AI startups within this ecosystem. Intense competition and rapidly changing technology mean that only the most adaptive startups continue to attract funding and interest. Furthermore, the skills gap remains a pressing issue. Effective AI development requires a highly skilled workforce, and with demand for such talent outstripping supply, startups must be resourceful in attracting and retaining the best minds.

Despite these challenges, the startup ecosystem for AI remains incredibly vibrant and full of potential. The venture capital trends of today show a marked maturity compared to the wild, speculative bursts of investment seen in the tech booms of the past. This maturity reflects an understanding of AI's complex nature and a commitment to backing companies that not only have pioneering ideas but also possess the capacity to implement these ideas in significant and profitable ways.

Imagine an old-world bazaar bustling with traders of all sorts trying to sell their wares. Now, picture the modern equivalent in the AI sector, where intellectual property and innovative applications are the items up for grabs. This is the current state of the AI startup ecosystem—a hotbed of ingenuity, where the stakes are high, and the spoils are unpredictable yet potentially monumental. It's a thrilling space that continues to captivate the imagination of venture capitalists and entrepreneurs alike.

Case Studies

Diving into the dynamic realm of AI startups, the success stories are both inspiring and instructional. Consider a startup that emerged from the shadows with a novel AI-driven solution for healthcare diagnostics, reshaping how diseases are identified in their early stages. Another startup, thriving in the bustling fintech sector, leveraged machine learning algorithms to transform risk assessments, gaining the trust of major financial institutions worldwide. Then there's the tale of an AI entertainment company breaking barriers by creating hyper-personalised content that captivates audiences, potentially revolutionising how we consume media. These case studies illustrate not just the potential of visionary ideas but also highlight the challenges and triumphs encountered on the path to success. Investors and innovators alike can glean valuable insights from these narratives about the nuances of navigating this ever-evolving industry. These stories remind us that with the right blend of creativity, technology, and tenacity, AI startups can indeed forge paths that redefine industries.

Successful AI Companies In the vibrant world of artificial intelligence, success stories abound, showcasing how businesses harness AI's potential to revolutionise industries. Emerging from small startups into industry titans, these companies demonstrate the limitless possibilities when innovation meets execution. Let's delve into the fascinating journeys of some noteworthy AI companies that have not only thrived but have set benchmarks in the technological realm.

Consider DeepMind, a British AI startup, which has been a trailblazer in artificial intelligence research and application. Acquired by Google in 2014, DeepMind's mission is to "solve intelligence" and then use that to tackle the world's most pressing problems. Known for creating AlphaGo, an AI programme that defeated the world's best Go player, DeepMind showed the world that machines could outperform humans in complex cognitive tasks. This monumental achievement wasn't just a triumph for DeepMind, but a significant leap for AI as a discipline.

While DeepMind focuses on research, UiPath has taken strides in operational AI, particularly in the field of robotic process automation (RPA). Founded in Romania, UiPath provides enterprises with advanced software platforms that automate mundane, repetitive tasks. This not only enhances productivity but also allows human employees to focus on more strategic initiatives. Their approach to automation through AI has seen them grow exponentially, with their solutions now being integral to businesses worldwide, ranging from banking to healthcare.

A glance at the AI landscape would be incomplete without mentioning OpenAI, an American company founded to ensure that artificial general intelligence (AGI) benefits all of humanity. Through its work on AI models like GPT-3, OpenAI has showcased the power of AI to generate human-like text and foster human-computer interaction. The company's emphasis on ethical AI development is as noteworthy as its technological feats, inspiring others in the industry to ponder the moral responsibilities accompanying innovation.

In the ecommerce sector, Alibaba's ambition and innovation shine through. Their AI-driven logistics, customer service chatbots, and personalized shopping experiences set new standards in the retail industry. By using AI to analyze shopping habits, Alibaba tailors suggestions to individual consumers, improving user experience and driving sales. Their success story highlights how sector-specific AI solutions can transform traditional business models.

The healthcare sector too has seen its share of AI success through companies like Zebra Medical Vision. Based in Israel, Zebra Medical Vision has developed AI algorithms capable of reading medical imaging, thus accelerating diagnoses and highlighting potential abnormalities. Their AI tools have been pivotal in early detection of diseases like cancer, saving lives and reducing the burden on healthcare professionals. Zebra's success underscores the life-changing impact AI can have when applied thoughtfully to healthcare.

In the same spirit, Lemonade has been redefining the insurance industry. Using AI and behavioural economics, Lemonade offers a refreshingly new take on insurance policies. Its platform uses chatbots and machine learning algorithms to process claims swiftly, often in minutes. By minimising the bureaucratic delays associated with traditional insurance, Lemonade provides a seamless and transparent customer experience. This innovative approach has disrupted the insurance market, making AI a cornerstone of their business strategy.

Beyond individual companies, AI's success can also be measured by platforms like Nvidia. Known for its graphics processing units (GPUs) that initially targeted gaming, Nvidia quickly capitalised on the potential of GPUs in AI. Today, their hardware accelerates AI applications across industries, from deep learning to self-driving cars. Nvidia's journey showcases how adaptable, dual-focus strategies can turn industry trends into profitable opportunities.

As we navigate these narratives, one thing becomes abundantly clear: successful AI companies don't just innovate—they redefine. Their stories provide lessons not only in technology but also in adaptability, foresight, and responsibility. They remind us of AI's transformative power and the boundless possibilities it offers when paired with human ingenuity.

The barriers each of these companies overcame weren't just technical. They had to instil trust, address ethical concerns, and navigate regulatory landscapes. This multifaceted success highlights the importance of holistic approaches where societal impact and business acumen go hand in hand. These companies did not simply ride the technological wave; they crafted it, shaping industries as they went.

The path to AI success is as varied as the sectors these companies operate in—a testament to AI's versatility. By capturing and analysing massive datasets, these enterprises create value in unforeseen ways. However, it's crucial to remember that behind algorithms are human minds driving creativity and strategic thinking. By fostering a culture of innovation and encouraging a blend of multidisciplinary collaboration, successful AI companies remain at the forefront of the digital age.

Looking forward, the inspiration derived from these AI juggernauts can fuel new dreams and potential enterprises. As more players enter the field, they too can glean insights from the successes of their predecessors. The landscape of AI is expansive and constantly evolving, presenting countless unexplored opportunities. Those ready to engage with these challenges will find that, like their predecessors, the potential rewards are substantial.

Thus, in examining successful AI companies, we not only celebrate achievements but also set a benchmark for what is possible. These companies are beacons of what's achieved at the intersection of science, entrepreneurship, and societal needs. As torchbearers in AI, they're guiding the industry into uncharted territories, challenging us all to think bigger and better about what's possible in a world brimming with technological promise.

CHAPTER 24: LEARNING AI: RESOURCES AND TOOLS

Venturing into the world of artificial intelligence can be both inspiring and daunting, but fortunately, there's an ever-growing array of resources and tools available to guide one's journey. Online courses and tutorials offer structured learning from the comfort of one's home, with platforms like Coursera, Udacity, and edX presenting opportunities to delve into AI and machine learning at various skill levels. For those who prefer diving into code, open-source projects such as TensorFlow and PyTorch provide hands-on experience and a welcoming way to get one's hands dirty with real-world applications. The open-source community thrives on collaboration and shared learning, encouraging enthusiasts to contribute and learn alongside seasoned experts. This abundance of resources empowers both novices and professionals to learn at their own pace, ensuring a comprehensive understanding of AI's incredible potential and its role in shaping our world. By piecing together these valuable educational tools, learners can build a robust foundation in AI, balancing theoretical knowledge with practical application, and truly become part of this cutting-edge field's vibrant future.

Online Courses and Tutorials

The digital age has brought a wealth of opportunities for those eager to dive into the world of artificial intelligence, and online courses and tutorials stand out as some of the most accessible paths. Whether you're a seasoned technology professional or a curious enthusiast, these resources offer a treasure trove of knowledge that's only a click away. Platforms like Coursera, edX, and Udacity have collaborated with leading universities and AI experts to create comprehensive courses that cover everything from basic AI principles to advanced machine learning techniques. As you engage with these materials, you'll find that each course is designed not just to inform, but to inspire and challenge your understanding of what's possible with AI today. It's a space where the boundaries of human intelligence and machine capabilities blur, and creativity finds new expressions. These courses don't just teach you how to use AI tools; they equip you with the skills to innovate and adapt in a rapidly evolving landscape, empowering you to take part in shaping the future of technology. With an open mind and a bit of perseverance, these tutorials can transform your AI curiosity into a profound journey of discovery and innovation.

Educational Platforms have truly democratized the learning landscape, opening doors to knowledge that transcend traditional educational confines. For those delving into the world of AI, selecting the right platform can make the difference between superficial understanding and deep, meaningful insight. The digital sphere is bustling with choices, each offering a distinct learning experience, curated to meet various learner needs and expertise levels.

One of the flagship platforms in this domain is Coursera. It collaborates with top universities and organisations worldwide to deliver courses that cater to various skill levels—from beginner introductions to advanced machine learning applications. By providing interactive quizzes, peer-graded assignments, and its hallmark capstone projects, Coursera enables learners to apply theoretical knowledge in practicable ways. This not only enriches the user's experience but also ensures they emerge from their courses with demonstrable skills.

Similarly, edX is a noteworthy mention, founded by Harvard and MIT. Like Coursera, it partners with leading universities to offer courses rich in content and structured in a comprehensible manner. It's not just about taking a course but immersing oneself in an environment that's rigorous and stimulating. Furthermore, edX's MicroMasters programs are designed to offer a bridge to advanced education and even count towards a full master's degree at some institutions. The platform also empowers learners by offering a flexible schedule, crucial for those balancing education with other commitments.

For the more technically inclined, platforms like Udacity have carved out a niche. Born from a Stanford experiment, Udacity's Nanodegree programs are hands-on and focus heavily on projects that mirror real-world scenarios. This skills-first approach is particularly beneficial in AI, where theory swiftly turns into practice. What sets Udacity apart is its collaboration with industry leaders such as Google and IBM, which keeps the content relevant and updated with current technological trends. By focusing on career preparedness, Udacity ensures that its learners are industry-ready upon course completion.

Meanwhile, DataCamp targets those keen on data science and analytics, essential components of AI. Its user-friendly interface, coupled with a focus on coding exercises, makes it an excellent choice for beginners. The learning paths are streamlined to avoid overwhelming newcomers, yet comprehensive enough to cater to intermediate and advanced users. DataCamp supports learning through doing, a philosophy it holds dear, which is evident from its interactive coding exercises and real-life case studies that develop not just knowledge but applicable skills.

Another platform, FutureLearn, emphasises a community-centric approach to learning. Ideal for those who thrive in collaborative environments, FutureLearn's AI courses foster interaction through discussion forums and group-based projects. This is particularly beneficial in areas of AI where problem-solving is often derived from diverse perspectives and collaborative ideation. By contextualising learning within a social framework, the platform enhances the engagement level of its offerings.

In juxtaposition to these structured, often formal platforms, there is also a refreshing mix of creative, informal learning spaces like Khan Academy and YouTube. While traditionally known for mathematics and general education, Khan Academy's foray into AI basics provides a gentle introduction to key concepts with animated explanations and easy-to-follow narratives. YouTube, on the other hand, is a treasure trove of tutorials, with educators and experts alike offering visual lessons that vary in depth. The advantage here is the immense variety and viewpoints, though it requires learners to be discerning about the content's credibility.

In another vein, GitHub isn't a traditional educational platform but stands as a self-directed learning hub. With countless AI projects and open-source repositories, it provides learners a sandbox of opportunity to explore, modify, and learn from existing codebases. There's a profound understanding that comes from dissecting and enhancing real projects, offering insights that transcend built-in platform tutorials. Contributors often follow through detailed documentation, providing an informal yet robust teaching resource.

Finally, for those inclined towards free resources, platforms like Stanford Online and MIT OpenCourseWare grant public access to some of their pioneering AI courses. These resources level the playing field, offering high-quality content without the fee barrier, making cutting-edge AI education accessible globally. However, learners may find themselves navigating these resources without the structured support systems present in other platforms, requiring more self-discipline and initiative.

As the AI field continues to expand and evolve, the line between formal education and independent learning blurs. The proliferation of educational platforms equips anyone with curiosity and a bit of tenacity to acquire knowledge at their own pace. Whether considering a full-scale career transformation or simply broadening one's intellectual horizons, these platforms serve as critical gateways in the quest for understanding artificial intelligence.

Open Source Projects

Dive into the world of open source projects and you'll discover a landscape brimming with opportunities to learn and create. These projects serve as the breeding ground for innovation, inviting enthusiasts and seasoned developers alike to contribute and collaborate. As you're tinkering with lines of code, you become part of a dynamic ecosystem that nurtures skills and fosters creativity. From libraries like TensorFlow to platforms such as PyTorch, open source initiatives provide the essential building blocks for exploring complex AI concepts. More than just coding, it's about joining a global community dedicated to pushing the boundaries of possibility in AI. By contributing to these projects, you enhance your skills and potentially make a mark on the technological world, driving the next wave of artificial intelligence advancements while learning from pioneers along the way.

Hands-on Learning is where you truly start to roll up your sleeves and immerse yourself in the world of artificial intelligence. Many believe that theory lays the foundation, but it's the practical application that really sets your understanding of AI apart. Whether you're a tech enthusiast eager to experiment, a professional looking to deepen your expertise, or someone simply curious about the field, open source projects offer an invaluable resource for hands-on learning. They not only provide the opportunity to apply what you've learned but also expose you to real-world challenges and solutions that theoretical studies might not cover.

Open source projects act as a bridge between theoretical understanding and practical skills. They offer a playground for experimenting with AI tools and resources without the prohibitive costs often associated with proprietary software. Imagine having access to a vast catalogue of projects where you can observe AI in action, tweak it, retrain it, and even contribute to its development. This is the beauty of open source. Not only do these projects enhance your technical skills, but they also foster a sense of community as enthusiasts and professionals collaborate, share ideas, and push the boundaries of what AI can achieve.

There are numerous open source platforms that invite you into the fray. For instance, TensorFlow and PyTorch are not just buzzwords in the AI community. These libraries, supported by Google and Facebook respectively, are comprehensive ecosystems for building machine learning models from scratch. By engaging with these platforms, you gain insight into the inner workings of neural networks, from perceptrons to deep learning behemoths. You learn by doing, adopting an iterative approach to solving complex problems, which is far more rewarding than any textbook explanation.

Venturing into open source AI projects also means you're part of a larger movement towards democratizing AI. Tools like FastAI build on existing libraries by simplifying the learning process, making AI more accessible to those without a specialised background. These tools come with tutorials, pre-trained models, and user-friendly interfaces that guide you through the convolutions of deep learning. Taking part in such projects demystifies the complexities of AI and presents you with the opportunity to use it in everyday applications, whether it's in image recognition, language processing, or even crafting clever chatbots.

Participation in open source projects often begins with forking a repository or downloading a project and setting up a development environment. This hands-on approach teaches not just the in-depth subject matter but also soft skills that are crucial in technology careers today. Problem-solving, critical thinking, and collaboration are just a few of the competencies you'll hone. As you work through challenges, there's a thrill in debugging a problem and the satisfaction of contributing code that others might use or build upon. It's this cycle of learning, creating, sharing, and improving that makes the open source community vibrant and inspiring.

Furthermore, immersing yourself in open source AI projects allows you to witness firsthand how ideas transition from conceptual frameworks to robust applications. Take, for instance, the development of recommendation engines. By engaging with projects that focus on personalisation algorithms, you see how complex datasets are analysed and learn the intricacies of tailoring user experiences in various domains, from e-commerce to streaming services. These insights make you not merely a consumer of technology but a contributor to its evolution.

An added advantage of engaging in open source projects is the portfolio you can build. Real-world hands-on experience is something prospective employers highly value. By contributing to projects, troubleshooting issues, and collaborating with other developers, your GitHub profile for example becomes a living portfolio of your skills and achievements. It doesn't just showcase your technical abilities but tells a story of your problem-solving capabilities and your commitment to continuous learning.

Diving into open source projects isn't without its challenges. It requires patience and resilience. There will be moments of frustration as you tackle steep learning curves or experience code that just won't compile. However, it's these very challenges that strengthen your understanding and appreciation of AI. They teach you to be resourceful, to seek out documentation, forums, or communities where you can find solutions or even contribute to future documentation with your newfound knowledge.

There's something profoundly satisfying about seeing your contributions live in a project used by others across the globe.

It's a testament to the collaborative nature that open source embodies, turning individual efforts into collective advancements. As you engage with these projects, you also gain exposure to the diversity of approaches within the AI space. You'll learn that there's often more than one way to solve a problem, and this multiplicity of perspectives enriches your own problem-solving toolkit.

The dynamic field of AI continues to expand its reach, and staying current means being proactive, continuously honing your skills. Engaging in open source projects in AI is a wonderful way to do this. It ensures that you are not only consuming technology but actively interacting with it, exploring its boundaries, and perhaps even redefining them. AI's future is in the hands of those willing to learn, to experiment, and to share. In this journey, hands-on learning through open source projects stands as a testament to the promising horizon of AI and your place in shaping it.

CHAPTER 25: REAL-WORLD AI APPLICATIONS

In today's fast-paced digital age, artificial intelligence is no longer just a futuristic concept; it's deeply interwoven into the fabric of diverse industries, forging new pathways and redefining what is possible. From healthcare to finance and manufacturing to entertainment, AI is the powerhouse behind transformative technologies that revolutionise our day-to-day experiences. Imagine intelligent algorithms supporting doctors in diagnosing diseases with precision or crafting personalised playlists that perfectly match your mood. Companies have begun to implement AI-driven insights for streamlining operations, hinting at a future where efficiency and innovation go hand in hand. Looking at real-world case studies reveals a mosaic of AI's potential, tinged with lessons learned along the way—such as the vital need for ethical considerations and robust data governance. These lessons guide industries in adopting best practices, ensuring AI not only augments capabilities but also aligns with societal values, fostering a collaborative future shaped by technological progress.

Case Studies

In the fascinating realm of real-world AI applications, case studies serve as compelling narratives of innovation, adaptation, and success. These snapshots offer a deep dive into how various industries creatively leverage AI to revolutionise processes, enhance productivity, and tackle unique challenges. From the bustling corridors of healthcare facilities using AI diagnostics to enhance patient care, to financial institutions employing algorithmic trading systems that redefine market engagement, the breadth of AI is astonishing. In manufacturing, smart factories utilise robotics and automation, setting the stage for a new industrial revolution. Each case becomes a testament to AI's transformative power, highlighting custom solutions tailored to industry-specific needs. These stories not only illuminate best practices but also inspire future developments, illustrating the potential of AI to reshape the fabric of everyday life. As we delve into these studies, we gain insights into the common threads of ingenuity and perseverance that drive AI's expansive influence across global sectors.

Industry-Specific Implementations delve into how artificial intelligence is reshaping various sectors through tailored applications. By examining specific examples within these industries, we can gain a deeper understanding of the transformative power AI holds. It's not just about the technology itself, but rather how it's being woven into the fabric of these fields to solve real-world problems in innovative ways.

In agriculture, AI has begun to play a pivotal role in precision farming. Farmers now use machine learning algorithms to analyse vast datasets, optimising crop yields and managing resources more efficiently. For instance, AI-powered drones survey extensive fields, collecting data that's processed to inform decisions about watering, fertilising, and harvesting. By reducing resource waste and improving crop production, AI is not only enhancing productivity but also contributing to sustainable farming practices, which are crucial in a world facing climate change and population growth.

In the healthcare domain, AI's impact is profound and multifaceted. From diagnostics to patient care, AI technologies are being implemented to improve outcomes and efficiency. One standout example is the use of AI in radiology, where image recognition algorithms assist in detecting anomalies in medical images with remarkable accuracy. This capability aids radiologists by providing second opinions and highlighting areas of concern more swiftly than traditional methods. It's a prime instance of AI augmenting human expertise, leading to faster diagnosis and treatment, which can be life-saving.

The financial sector has embraced AI in risk management, fraud detection, and customer service to streamline operations and enhance security. Algorithms power real-time transaction monitoring systems that detect fraudulent activities by identifying unusual patterns and alerting authorities swiftly. This automation not only curtails financial crime but also optimises resources by reducing the need for extensive manual checks. In banking, chatbots and virtual assistants have improved customer engagement by providing quick, accurate responses to queries, available round the clock, thus enhancing customer experience significantly.

Manufacturing industries are witnessing a revolution driven by what is often termed Industry 4.0, where AI and the Internet of Things (IoT) converge. In smart factories, AI enables predictive maintenance, whereby machinery performance data is continuously monitored to predict failures before they occur, thereby minimising downtime. These AI systems learn from historical data, optimising production processes and ensuring consistent quality. Furthermore, advanced robotics powered by AI are capable of handling complex assembly tasks that were once the realm of human workers, increasing precision and reducing errors.

Retail businesses are leveraging AI to revolutionise the shopping experience with personalisation and operational efficiency. Online platforms use recommendation engines to analyse consumer behaviour and provide tailored product suggestions, enhancing customer satisfaction and increasing sales. Moreover, inventory management systems employ AI to predict demand, which optimises stock levels and reduces waste. In physical stores, AI technologies enable cashier-less checkouts and dynamic pricing, furthering convenience for consumers and efficiency for store operators.

The energy sector has also seen innovative AI applications aimed at optimising operations and supporting renewable energy integration. AI algorithms manage grid stability by forecasting supply and demand fluctuations, allowing energy companies to adjust production accordingly. This ability is particularly critical as more renewable energy sources, such as wind and solar, are added to the grid, which traditionally has been less predictable due to weather dependency. AI-driven energy management systems ensure a balanced supply, reducing reliance on fossil fuels and promoting a sustainable energy future.

In entertainment, AI's role is both fascinating and controversial. From generating music to creating digital art, AI algorithms are pushing the boundaries of creativity. Streaming services employ AI to enhance content delivery through personalised recommendations, while AI-generated scripts and virtual actors are starting to emerge in the film industry. However, these advancements also raise ethical questions about authorship and originality. Despite the debates, AI's ability to innovate and inspire in the creative industries is undeniable.

The education sector is experiencing paradigm shifts with AI's introduction as a tool for personalising learning experiences. Adaptive learning systems customise educational content to fit individual student needs, promoting better engagement and understanding. These systems use AI to assess student performance in real-time, offering additional resources or alternative approaches as required. What emerges is a more inclusive and effective learning environment where educators can focus on areas needing human intervention, catalysing the potential for each learner.

Transportation systems, enhanced by AI, are transforming urban life with applications like traffic management and autonomous driving. Cities around the world are deploying AI-powered traffic control systems to reduce congestion and improve safety. These systems analyse traffic data to optimise light signalling, reroute flows, and even predict traffic patterns. Furthermore, the integration of AI in public transit systems provides predictive maintenance and service scheduling, enhancing reliability and efficiency for commuters.

Finally, AI technology is integral to breakthroughs in the field of space exploration. Autonomous systems guide exploratory missions, enabling remote robotic operations on distant planets. AI analyses vast amounts of data gathered by satellites and telescopes to reveal insights that may otherwise remain hidden. From mapping the Martian surface to discovering distant galaxies, AI assists scientists in pushing the boundaries of human knowledge about the universe, making space exploration more attainable and cost-effective.

Each of these industry-specific implementations shows that AI isn't just a set of powerful algorithms. It's a catalyst for innovation, providing industries with tools that transform standard operations into sophisticated processes. This transformation is a peek into a future where industries will be more efficient, personalised, and aligned with sustainable practices. As we continue to explore these implementations, the potential for AI to redefine societal norms becomes increasingly apparent, making it an exciting era for technology enthusiasts and professionals alike.

Lessons Learned

As we delve into the intricacies of "Real-World AI Applications", it's clear that AI's integration into everyday life is both a thrilling and complex dance. The lessons learned from its ongoing deployment across industries are as varied as the sectors themselves. Adaptability emerges as a critical trait, as each application often demands custom solutions and the willingness to evolve with unforeseen challenges. Human oversight remains indispensable, serving as a crucial check on AI's automated processes, thus ensuring ethical integrity and practicality align. Moreover, cross-disciplinary collaboration has proven invaluable, where technologists, industry experts, and ethicists come together to ensure AI's benefits are maximised responsibly. These insights illuminate a path forward, where learning from both triumphs and missteps will shape the more efficient and equitable future AI promises to deliver.

Best Practices in the domain of Real-World AI Applications are crafted by distilling the "Lessons Learned" from various industries. These practices aren't just platitudes delivered from the high towers of academia; they're grounded in the practical challenges that organisations face daily. As AI continues to evolve, being methodical about what works, what doesn't, and how it can be improved is critical for the technology's sustainable growth.

AI applications have permeated nearly every facet of modern life, revolutionising fields from healthcare to finance, and beyond. Yet, uniform success isn't always guaranteed. Thus, capturing "Lessons Learned" offers a treasure trove of insights for refining operations in this space. The first and possibly the most significant lesson learned is that data is the lifeblood of AI. Poor data results in poor models, irrespective of the algorithms used. Therefore, companies have realised the importance of investing in robust data collection and cleaning mechanisms. A staggering amount of effort is poured into what might seem mundane tasks, like data labelling and bias correction, but this work pays off in producing more reliable AI systems.

In the rush to implement AI solutions, organisations have also learned the value of starting with smaller, manageable projects. Going for grandiose projects at the outset might be tempting, yet it often leads to failure due to unforeseen complexities. Thus, a best practice emerged: begin with pilot projects that help to uncover any hidden variables or challenges. These smaller projects allow a company to iterate and refine before rolling out on a larger scale.

Moreover, transparency in AI operations has proved to be essential. People naturally distrust what they don't understand, so deploying 'black box' AI systems often meets with resistance both internally and externally. To counter this, explainable AI is becoming a mainstay in real-world applications. Indeed, transparency isn't just an ethical concern—clarity in operations can enhance user trust and pave the way for smoother implementations. AI systems that offer justifications for their decisions tend to be more readily accepted, especially in sensitive fields like healthcare or finance.

Next up is collaboration. Looking back, one can see that the successful deployment of AI often involves strong interdepartmental teamwork. It's no longer just the purview of data scientists locked away in silos; rather, successful AI projects draw on a range of insights, from domain experts to end-users. This synergy enriches the AI development process, as real-world insights can guide how data models are structured. Collaborations also pave the way for multi-disciplinary innovation, unlocking creative solutions to complex problems.

Furthermore, continuous monitoring and updates are essential in AI applications. AI systems do not operate in static environments; they are nested in dynamic, ever-changing landscapes. Thus, treating AI systems as 'set-and-forget' tools is a recipe for obsolescence. The best practices include regularly updating models based on new data and monitoring for any unintended biases that may inadvertently have crept in. This need for vigilance points to the importance of feedback loops in all AI applications, allowing systems to adapt and evolve in the face of evolving needs and challenges.

Security and ethical considerations form another cornerstone of best practices. With data platforms being the new frontier for cyber threats, ensuring robust data security protocols is not only a technical requirement but also a legal and ethical imperative. The implementation of privacy-by-design principles, ensuring anonymisation where required, and thoughtful consent management are prime examples of how lessons from past missteps have resulted in clearer compliance and ethical guidelines.

Another valuable lesson is recognising the limits of AI. As powerful as these systems can be, they aren't infallible and require human oversight. Human-machine collaboration has underscored this point time and again. AI should augment human decision-making, not replace it entirely. Hence, designing AI systems that accommodate human insight and judgment often leads to better outcomes. This approach not only enhances efficiency but also helps capture aspects that purely quantitative measures might overlook.

Organisations have also learned to manage expectations effectively. Overhyping AI's capabilities can lead to disillusionment and mistrust. It is critical to set realistic goals and timelines. The best practices in the field involve transparent communication about the potential scope and limitation of AI technologies, both within an organisation and to external stakeholders.

Finally, embracing a culture of innovation and learning is paramount. AI technology is still relatively nascent and fast-evolving. Keeping abreast of the latest developments ensures that companies remain competitive and can leverage cutting-edge solutions. This often involves investing in continuous training for employees, supporting in-house talent to explore new methodologies, and encouraging intellectual curiosity to discover new potential applications and solutions.

In summary, the journey of integrating AI into real-world applications is one of constant learning and adaptation. By adhering to the best practices derived from lessons learned, organisations position themselves to not only meet the challenges of today but also to build resilient frameworks for the future. As AI continues its march into the layers of everyday life, these learnings and best practices chart a course for responsible and effective implementation.

CONCLUSION

As we find ourselves at the precipice of a new technological era, the journey through the landscapes of artificial intelligence leaves us both enlightened and contemplative. We've unravelled the intricacies of AI's evolution, explored its application across various domains, and delved into the ethical conundrums it presents. From healthcare to space exploration, AI has proven to be a catalyst for innovation, pushing the boundaries of what's possible. Yet, the more we advance, the essential it becomes to tread with caution, ensuring that this potent tool serves humanity ethically and responsibly. The future of AI holds immense promise, but it demands our vigilance and foresight—an unwavering commitment to harnessing its power for the greater good. As we march forward, the harmonious coexistence of AI and humanity must be at the forefront of our endeavours, fuelling a future where technology not only enhances but also respects the essence of our human experience.

Recap of Key Points

As we draw to a close, it's important to reflect on the central themes that have navigated our exploration of artificial intelligence. AI is woven deeply into the fabric of our modern society, not just as a tool for innovation but as a transformative force reshaping industries and everyday life. The journey from understanding AI's basic concepts to envisioning the future of artificial general intelligence is an odyssey of imagination as much as it is of technology.

At the heart of AI lies the dynamic interplay between machine learning and deep learning. These technological pillars underscore much of AI's potential, unlocking capabilities that allow machines to learn from data, identify patterns, and make decisions with increased sophistication. The evolvement of algorithms and models, coupled with neural networks' rapid advancements, has widened AI's scope, reaching into domains previously confined to human intelligence alone.

Moreover, AI's impact stretches across various sectors, creating efficiencies and fostering new possibilities. In healthcare, for example, AI is revolutionising diagnostic tools and promoting personalised medicine, with life-altering implications. Finance, too, is witnessing AI's prowess through algorithmic trading and cutting-edge fraud detection mechanisms, enhancing security and precision in a domain inherent with risks.

The automotive industry is another exemplification of AI's transformative power. Self-driving technologies are not just theoretical but are gradually becoming a reality, overcoming challenges and sparking legal and ethical debates along the way. Manufacturing sees similar disruptions, with smart factories and predictive maintenance enhancing productivity and efficiency at unprecedented levels.

But AI isn't only about efficiency; it's also about creativity. Whether in AI-generated content for entertainment or in redefining art and literature, we've glimpsed how AI can push the boundaries of what is creatively possible. The philosophical questions arising from AI's creative and seemingly intelligent behaviour invite us to reconsider the very nature of intelligence and consciousness.

Ethical considerations form a critical axis around which AI discussions revolve. Addressing issues like bias, privacy, and employment displacement remains crucial as AI continues to permeate different social strata, pushing us to ponder on responsible and inclusive AI governance. The role of AI in society isn't confined to technological advances but extends into deeper, philosophical dialogues about our future coexistence with intelligent machines.

Finally, our exploration circles back to a core moral imperative: the future directions of AI. The landscape ahead is abundant with both promise and peril. As we look forward, encouraging responsible AI development becomes not merely desirable but necessary. Nurturing governance structures and cultivating collaborative human-AI relationships will be vital in steering AI towards improving lives and solving complex global challenges.

Future Directions

As we stand on the precipice of profound technological shifts, the future directions of AI beckon with both promise and introspection. The path forward is not just about unprecedented innovations but also about harmonising these advancements with human ethics and societal wellbeing. AI's potential to transform industries, redefine job markets, and reshape global economies presents a canvas of opportunities as well as challenges. We must navigate this future by fostering a culture of responsible AI development that prioritises transparency and accountability. Our journey must embrace a holistic approach that melds technical prowess with a commitment to human values. The dialogue between technologists, policymakers, and society needs to be continuous and inclusive, ensuring that AI serves as an enabler of progress that aligns with collective aspirations. Such synergy will be crucial in crafting a sustainable future where artificial intelligence complements, rather than competes with, our inherently human creativity and wisdom.

Encouraging Responsible AI Development in the domain of future directions requires us to critically examine how our innovations align with long-term societal goals. The potential of artificial intelligence, vast and undisputed, brings with it a tremendous responsibility. As technology enthusiasts and professionals, we must ensure that the AI systems we create are not only robust and efficient but also ethical and beneficial to humanity. There's no denying that AI has the power to transform industries and reshape our daily lives, yet with this power comes the duty to steer these advancements towards the greater good.

The conversation around responsible AI development isn't new, but it's gaining momentum as AI technologies become more ingrained in our lives. Developers, researchers, and policymakers are increasingly acknowledging the need for AI systems that reflect human values and ethics. The challenge lies not just in building intelligent systems but in ensuring they operate within the bounds of fairness, transparency, and accountability. This isn't merely a technological issue; it's a societal imperative.

To navigate this complex landscape, interdisciplinary collaborations are crucial. Bringing together experts from diverse fields, including ethics, law, sociology, and computer science, can help guide the equitable development of AI. By fostering dialogue and understanding across these disciplines, we can cultivate an environment where moral guidelines evolve alongside technological innovations. Together, these insights can guide policy-making and the establishment of industry standards that ensure AI systems don't inadvertently perpetuate biases or inequalities.

Effective regulation is another pillar of responsible AI development. Governments and international bodies must work toward creating a framework that balances innovation with protection. Crafting policies that encourage innovation while safeguarding public interests is a delicate task, but necessary. Regulations must be flexible enough to adapt to technological advances yet robust enough to prevent misuse.

In this context, having a global perspective is vital, recognising that AI operates beyond borders and cultures, impacting a global populace.

Another aspect of encouraging responsible development is through education and outreach. It's imperative that we equip the next generation of AI professionals with not just technical skills but also a strong ethical foundation. Educational programmes should integrate ethics and social responsibility into their curriculums. Engaging with the public to demystify AI and highlight its potential impacts can foster informed discussions and a more educated populace that can actively engage in shaping AI's future.

Incorporating ethical AI into corporate culture is equally important. Companies should take proactive steps in addressing the ethical implications of their AI solutions. This involves setting up internal guidelines and review processes that prioritise ethical considerations throughout the AI lifecycle from conception to deployment. Such initiatives should encourage transparency and allow for accountability, ensuring that AI products can stand up to public scrutiny without compromising on ethical standards.

Moreover, establishing platforms for public engagement in AI discourse is beneficial in building trust and understanding. When citizens have a say in how AI technologies are implemented, it builds a shared sense of responsibility and ownership. Public consultations, AI ethics panels, and scenario planning workshops are effective tools for incorporating diverse viewpoints and highlighting real-world concerns that may otherwise be overlooked in tech-centric discussions.

The journey to responsible AI development also hinges on continuous evaluation and adaptation. As AI technologies evolve, so too must our understanding and frameworks of responsibility. This involves regular reassessment of AI systems to ensure they are delivering on ethical commitments. Open lines of communication between developers, users, and policymakers can facilitate ongoing feedback loops, ensuring that ethical considerations are as dynamic as the technology itself.

In an age where AI determines the fate of economies and societies, overlooking its ethical considerations isn't just risky—it's dangerous. Yet, the challenge lies not merely in recognising these responsibilities but in addressing them comprehensively. Envisioning future directions with responsible AI at the helm calls for a foundational shift in how we perceive the role of technology in society. It's about envisioning a future where AI not only anticipates human needs but also aligns with our core ethical values, ensuring progress that is equitable and sustainable. Let's make this vision of responsible AI development not just a future direction but a present priority.

APPENDIX A: APPENDIX

As we draw the curtains on our exploration of the vast and intricate world of artificial intelligence, this appendix serves as a complementary guide, filling in any gaps and providing additional context where needed. It's like the footnotes of a thrilling mystery, revealing the background to the layers we've peeled away throughout this journey. While technology enthusiasts and professionals may have found the chapters engrossing, this section is more of a breather, offering perspectives and clarifications on any potential ambiguities left in the wake of our discussions.

Often, within the realm of AI, there's a plethora of terms, methodologies, and concepts that can feel overwhelming. Here, we offer a concise but detailed account of any crucial details that didn't find their place in the main chapters—a nod to the brilliant minds who've driven AI's formidable rise.

In unpacking the technicalities, imagine this appendix as an echo of the conversations you might have had within the main content, the bits that linger in your mind long after the last page has been turned. It's a moment to step back and absorb the breadth of artificial intelligence's capabilities, its untapped potential, and the ethical dimensions that accompany its evolution.

We've shared glimpses of AI's impact across various sectors—from revolutionising healthcare to reshaping educational landscapes. However, the journey doesn't end here. The appendix acts as a compass guiding further inquiry and understanding, pointing you towards additional topics ripe for exploration.

Remember, this is not merely a summary but a launching pad for future curiosity—an acknowledgment that within the appendix lies the opportunity to ponder, reflect, and consider the nuances that make AI both a marvel and a matter of profound responsibility. So, let's embrace this appendix not just as an endnote, but as a catalyst for ongoing dialogue and discovery in the world of artificial intelligence.

- Additional clarifications on AI methodologies
- Greater understanding of complex AI topics
- Further reflections on ethical considerations
- Guidance for future learning and exploration

This is where the groundwork from the preceding chapters and personal exploration blends, paving the way for the next chapter in your AI journey. Consider it a toolbox, filled with insights and perspectives that might not change the world overnight, but just might change yours.

GLOSSARY OF TERMS

Diving into the world of artificial intelligence requires a grasp of the essential terminology that defines the field. This glossary aims to clarify and inspire curiosity as you explore the potential and intricacies of AI.

- **Algorithm:** A set of instructions or rules designed to perform a specific task or solve a problem. In AI, algorithms are the backbone of machine learning and decision-making processes.

- **Artificial General Intelligence (AGI):** The capability of a machine to perform any intellectual task that a human can do, representing the apex of AI development where machines exhibit flexible cognitive abilities.

- **Autonomous Vehicle:** A self-driving car or vehicle equipped with AI systems that allow it to navigate and operate without human intervention.

- **Bias:** The tendency of an AI system to produce prejudiced outcomes due to flawed data or algorithm design. Bias is a significant ethical concern in AI development.

- **Deep Learning:** A subset of machine learning characterised by neural networks with many layers, enabling the processing of large amounts of structured or unstructured data for tasks like image and speech recognition.

- **Ethics of AI:** The study and examination of moral, societal, and legal implications of AI technologies, focussing on issues such as fairness, accountability, and transparency.

- **Machine Learning:** A branch of AI focused on building systems that learn from data and improve their performance over time without being explicitly programmed for specific tasks.

- **Natural Language Processing (NLP):** A field of AI that deals with the interaction between computers and humans, enabling machines to understand, interpret, and respond to human language.

- **Neural Network:** A computational model inspired by the human brain, consisting of interconnected 'neurons' that process information in layers, used extensively in machine learning and deep learning.

- **Predictive Maintenance:** The use of AI to predict equipment failures before they occur, optimising maintenance schedules and reducing downtime.

- **Supervised Learning:** A type of machine learning where an algorithm is trained on labelled data, learning to map inputs to outputs to make predictions or decisions.

- **Unsupervised Learning:** A machine learning technique where algorithms identify patterns and structures in data without the need for labelled examples.

- **Virtual Assistant:** Software applications powered by AI that can perform tasks or provide services for individuals, usually via voice commands or text-based interactions.

These terms are just a starting point in understanding the vibrant world of AI. As we journey through the book, you'll see these concepts come to life, shaping the technologies of today and tomorrow.

Additional Reading

For those keen to delve deeper into the vast landscape of artificial intelligence, the sphere of additional reading offers a treasure trove of knowledge. Expanding your understanding of AI goes beyond grasping the glossary of terms. Consider books that provide historical context, like those tracing the evolution of AI from its early days to modern breakthroughs, offering insights into the minds that have shaped the field. Articles that explore current developments and debates are invaluable for staying abreast of ongoing technological transformations. Engaging with materials discussing ethical considerations can sharpen your awareness of the social and moral dimensions of AI. Whether your curiosity is piqued by machine learning intricacies or AI's role in reshaping industries, these recommended readings facilitate a more nuanced appreciation of AI's complexity and its unfolding impact on society.

Recommended Books and Articles When it comes to diving deep into the complex world of artificial intelligence, there are a few essential reads that stand out, forming a bedrock of understanding for enthusiasts and professionals alike. These resources, crafted by thought leaders, technologists, and theorists, each provide a unique perspective on AI's many facets, contributing valuable insights into its profound impact on society.

Start with "Superintelligence: Paths, Dangers, Strategies" by Nick Bostrom. It presents a compelling exploration of future scenarios where machines surpass human intelligence. The book delves into the existential risks associated with building intelligent machines and offers strategic frameworks for tackling those challenges. Bostrom's rigorous approach combines philosophy and technical expertise, making it an indispensable read for anyone interested in the ethical dimensions of AI.

Then there's "Life 3.0: Being Human in the Age of Artificial Intelligence" by Max Tegmark. This book takes readers on a journey through the potential futures AI may shape, examining both the utopian and dystopian possibilities. Tegmark discusses the philosophical and societal ramifications, encouraging readers to ponder what kind of future we want to build. His engaging narrative style makes complex ideas accessible, sparking inspiration and reflection.

For those drawn to the intersection of AI and creativity, "The Creativity Code: Art and Innovation in the Age of AI" by Marcus du Sautoy is a must-read. In it, du Sautoy explores whether AI can produce art that challenges our concept of creativity. He examines algorithms that are beginning to generate original art, music, and literature, raising questions about the nature of creativity itself. The book is not only informative but also invites readers to rethink what it means to be creative in an increasingly automated world.

Another critical read is "Prediction Machines: The Simple Economics of Artificial Intelligence" by Ajay Agrawal, Joshua Gans, and Avi Goldfarb. This book simplifies AI by breaking it down into its fundamental function—making predictions. It examines how AI advancements lower the cost of prediction and the cascading effects this has on decision-making processes across industries. With its focus on economic implications, the book offers practical insights for business leaders and policymakers looking to harness AI's potential responsibly.

Complement your reading with articles that offer fresh perspectives and updates on rapid developments within AI technology. Journals like "Artificial Intelligence" and "The Journal of Machine Learning Research" regularly publish pioneering studies that push the boundaries of what's possible. These peer-reviewed articles are invaluable for keeping up-to-date with the latest research findings and technological breakthroughs.

One can't overlook contributions from "MIT Technology Review" and "Wired", which frequently feature thought-provoking pieces about AI's intersection with various sectors such as healthcare, finance, and governance. These publications often include interviews with leading AI researchers and practitioners, providing firsthand insights into the challenges and innovations shaping the field.

Books such as "Deep Learning" by Ian Goodfellow, Yoshua Bengio, and Aaron Courville offer a deep dive into the technical aspects of machine learning. Their work is considered the definitive guide for those interested in the mechanics behind neural networks and deep learning techniques. While the material is dense, it is thoroughly rewarding for those keen on mastering the intricacies of AI.

Exploring the future, "The Master Algorithm: How the Quest for the Ultimate Learning Machine Will Remake Our World" by Pedro Domingos paints a picture of an algorithmic world. Domingos explores the five major schools of thought in machine learning, ultimately proposing potential solutions for achieving a master algorithm. The book invites readers to imagine a world radically transformed by such advancements, blending technical knowledge with visionary speculation.

Other articles worth noting come from sources like "The New York Times" and "The Atlantic," as these publications often cover AI's societal impact and ethical considerations. These pieces contribute to understanding how theoretical concepts are applied in real-world contexts, highlighting both benefits and unforeseen consequences.

If you're looking for an impactful historical perspective, Walter Isaacson's biographies, such as "The Innovators: How a Group of Hackers, Geniuses, and Geeks Created the Digital Revolution," offer valuable context. Though not exclusively about AI, these works provide understanding of the technological lineage that laid the groundwork for today's AI advancements. The book chronicles the people and innovations that have paved the way for our current tech landscape.

Lastly, for those who prefer concise overviews, Paul R. Daugherty and H. James Wilson's "Human + Machine: Reimagining Work in the Age of AI" presents a perspective on how AI is reshaping the workforce. The authors delve into the synergy between humans and machines, foreseeing a collaborative future where AI enhances human productivity. It's a hopeful take that presents AI not as a threat but as a tool for augmentation.

By immersing yourself in these recommended books and articles, you'll gain a profound understanding of the multifaceted world of AI. Each piece contributes uniquely to the dialogue, offering perspectives that challenge our assumptions about intelligence, creativity, and the future we are choosing to shape. Whether exploring ethical dilemmas or technical intricacies, these resources equip you with the knowledge to navigate the ever-evolving landscape of artificial intelligence.

Made in the USA
Monee, IL
04 November 2026

67375700R00154